UNIVERSITY OF STRATHCLYDE

30125 00518983 1

D1686063

J. VAGUE

C PE

ANDERSONIAN LIBRARY
★
WITHDRAWN
FROM
LIBRARY
STOCK
★
UNIVERSITY OF STRATHCLYDE

PLANNING, ESTIMATING, AND CONTROL OF CHEMICAL CONSTRUCTION PROJECTS

COST ENGINEERING

A Series of Reference Books and Textbooks

Editor

KENNETH K. HUMPHREYS, Ph.D.

Consulting Engineer
Granite Falls, North Carolina

1. Applied Cost Engineering, *Forrest D. Clark and A. B. Lorenzoni*
2. Basic Cost Engineering, *Kenneth K. Humphreys and Sidney Katell*
3. Applied Cost and Schedule Control, *James A. Bent*
4. Cost Engineering Management Techniques, *James H. Black*
5. Manufacturing Cost Engineering Handbook, *edited by Eric M. Malstrom*
6. Project and Cost Engineers' Handbook: Second Edition, Revised and Expanded, *edited by Kenneth K. Humphreys*
7. How to Keep Product Costs in Line, *Nathan Gutman*
8. Applied Cost Engineering: Second Edition, Revised and Expanded, *Forrest D. Clark and A. B. Lorenzoni*
9. Managing the Engineering and Construction of Small Projects: Practical Techniques for Planning, Estimating, Project Control, and Computer Applications, *Richard E. Westney*
10. Basic Cost Engineering: Second Edition, Revised and Expanded, *Kenneth K. Humphreys and Paul Wellman*
11. Cost Engineering in Printed Circuit Board Manufacturing, *Robert P. Hedden*
12. Construction Cost Engineering Handbook, *Anghel Patrascu*
13. Computerized Project Control, *Fulvio Drigani*
14. Cost Analysis for Capital Investment Decisions, *Hans J. Lang*
15. Computer-Organized Cost Engineering, *Gideon Samid*
16. Engineering Project Management, *Frederick L. Blanchard*
17. Computerized Management of Multiple Small Projects: Planning, Task and Resource Scheduling, Estimating, Design Optimization, and Project Control, *Richard E. Westney*
18. Estimating and Costing for the Metal Manufacturing Industries, *Robert C. Creese, M. Adithan, and B. S. Pabla*
19. Project and Cost Engineers' Handbook: Third Edition, Revised and Expanded, *edited by Kenneth K. Humphreys and Lloyd M. English*

20. Hazardous Waste Cost Control, *edited by Richard A. Selg*
21. Construction Materials Management, *George Stukhart*
22. Planning, Estimating, and Control of Chemical Estimation Projects, *Pablo F. Navarrete*
23. Precision Manufacturing Costing, *E. Ralph Sims, Jr.*

Additional Volumes in Preparation

Techniques for Capital Expenditure Analysis, *Henry C. Thorne and Julian A. Piekarski*

＃32018318

PLANNING, ESTIMATING, AND CONTROL OF CHEMICAL CONSTRUCTION PROJECTS

PABLO F. NAVARRETE
Independent Consultant
Cranford, New Jersey

Marcel Dekker, Inc.   New York•Basel•Hong Kong

Library of Congress Cataloging–in–Publication Data

Navarrete, Pablo F.
    Planning, estimating, and control of chemical construction
projects / Pablo F. Navarrete.
        p. cm. -- (Chemical industries ; 63) (Cost engineering ; 22)
    Includes bibliographical references (p. -) and index.
    ISBN 0–8247–9359–5 (hard cover : acid–free)
    1. Chemical plants--Design and construction. I. Title.
                              II. Series: Cost
engineering (Marcel Dekker, Inc.) ; 22.
TP155.5.N35   1995
690'.54--dc20   95–124
                                                      CIP

The publisher offers discounts on this book when ordered in bulk quantities. For more information, write to Special Sales/Professional Marketing at the address below.

This book is printed on acid-free paper.

Copyright © 1995 by MARCEL DEKKER, INC. All Rights Reserved.

Neither this book nor any part may be reproduced or transmitted in any form or by any means, electronic or mechanical, including photocopying, micro-filming, and recording, or by any information storage and retrieval system, without permission in writing from the publisher.

MARCEL DEKKER, INC.
270 Madison Avenue, New York, New York 10016

Current printing (last digit):
10 9 8 7 6 5 4 3 2

PRINTED IN THE UNITED STATES OF AMERICA

Margarita,

Thanks for your patience and love . . . .

FOREWORD

I wish that as a young engineer starting out in project management I had had this book. Experienced project managers had their collections of data, and one could always find books on the theory of project management; however, the data available was often inconsistent, and there was rarely an explanation of what it actually represented. Understanding the theory is great, but it does not give one the data and detail required to actually manage, estimate, schedule or contract.

In our contacts with plant project managers, we at FMC recognized that this informational void still existed and commissioned Pablo to fill it. This book represents the result of that effort. In addition to committing his knowledge to writing, Pablo has conducted many courses for our plant project managers and found that this material is just what they needed.

Although real data is a unique strength of this book, the text also provides theory and valuable guidance through checklists, *do's* and *don'ts*, and typical problems and corrective actions. As novice project managers plan, estimate, schedule, contract and actually execute their projects, they will understand what is important, the *whys* as well as the *whats*.

William K. Wakefield, Director
FMC Engineering Services
Princeton, New Jersey

PREFACE

An engineering project can be defined as a one-time endeavor to achieve finite quality, cost, and schedule objectives. When so defined the term applies to more than just the megaprojects initiated at a high corporate level and managed by experienced senior project managers supported by teams of project engineers and specialists, as well as purchasing agents. The term also applies to the small projects initiated by plant operations and/or maintenance and executed, practically single handed, by plant project engineers or, occasionally, by junior members of corporate engineering departments.

A vast majority of projects in the chemical industry fall under the last category. Yet, during my many years as a project engineer/manager reading books and going to seminars about project management and control, I have found that most of them are addressed to project and business managers. They deal mostly with the business and financial aspects of projects rather than with the actual execution details. They take much for granted and fail to delve into the *what* and *how* of the basic project legwork. It was always very difficult to find hard, practical data I could use in my day-to-day work.

I hope to fill the vacuum I perceive in the current project execution literature by offering project engineers and budding project managers a bird's-eye view of all facets of project execution (the *what*) and providing tools to perform the basic project legwork (the *how*).

This book is written for the hands-on project engineer, the guy in the trenches, the project roustabout, who "makes things happen" and insures smooth project execution, the guy who can make or break a project, without whom very few projects would be successfully completed.

Although this book is essentially a guideline for the execution and control of small projects, the concepts and techniques discussed can be applied to the various levels of project expertise and project complexity as well as the types of involvement in the project execution process. This book will be valuable to all those involved in the development, execution, and monitoring of chemical projects:

- For the **student** as well as the **novice project engineer/manager**, this book will provide the project ABC's.

- For the **managers** of small projects, it will be a tool box for their hands-on participation in all execution activities either by actual performance or checking the contractors performing them.

- For **seasoned project managers** in charge of major projects, it can be a refresher course in project management and a thorough checklist of project activities.   It also provides quick and easy ways to spot check the general contractor's work.

- The **supporting staff, business, production** and, especially, **venture managers** will find the material presented of the utmost interest.   The insight it provides on the execution of chemical projects will certainly enhance their performance and benefit all projects.

- **Engineering contractors** will also find the estimating tools presented in Chapter 13 very useful for the quick preparation of the preliminary estimates frequently submitted with proposals, as well as for checking their own definitive estimates.

This book reflects my personal observations, project execution practices, and opinions on how projects should be run. The procedures and guidelines presented were developed through the years to preserve and document those practices and ideas that not only proved to be accurate and time effective, but could also be implemented by simple manual means.

The proposed estimating and control procedures are not intended as substitutes for the detailed and sophisticated methods normally employed by contractors and propounded through the commercially available computerized programs for project management and control. They are intended to be tools to help the project engineer/manager evaluate project alternatives or monitor contractors' estimates and progress reports with limited in-house resources.

Like most project managers, I have been involved in and/or observed my share of bad projects. I am firmly convinced that most of the bad ones could have been avoided if these estimating and control procedures had been available and the recommended execution guidelines had been in effect.

I would like to share my experience with other project engineers/managers in the hope that doing so will contribute to promote their thinking, spur their creativity, enhance their performance, and result in a very high rate of successful projects.

In the summer of 1991, FMC Corporation asked me to prepare a Project Execution Guideline for the project engineers assigned as project managers for small projects that cannot justify the cost of a designated project team or a full-blown engineering and construction contractor.

A project manager working under those circumstances must be versatile and capable of personally executing many of the project tasks. To do so, the project manager needs tools. Thus, what was initially perceived as a rather simple guideline had to be complemented with a "tool box." For that purpose, I dug into my personal files for information compiled and/or developed through 30 years of project work, and updated and organized it for easy use by the project managers. The final product is not merely a guideline for the execution of small projects. It can also be a powerful tool for the project manager monitoring engineering and construction contractors executing projects of any size.

This is not a design book nor is it a cookbook. It is a guideline and thought provoker to point project managers/engineers in the right direction and promote creative thinking. They should become familiar with the proposed tools and view them not as ultimate goals, but rather as starting points to either modify them or develop new ones that reflect their own experience and address specific needs. To improve your own chance of success, it is important to remember that:

- Simple tools are the best tools.
- Every project is unique.
- There is no such thing as a recipe for successful project execution.

I want to express my gratitude to FMC Corporation and its Director of Engineering Services, Bill Wakefield, for their cooperation and patience during the last three years. In supporting the publication of this book, they have reaffirmed their commitment in the CICE guideline of sharing project execution information in order to enhance the performance of the entire construction industry.

Many friends and coworkers contributed in one way or another. Some like Nancy Buschman, Jack Gallagher, Jim Houle and Bill Wakefield offered very valuable comments and suggested changes. Others, like John Nabors, Manny Oconer and Dick Troell, allowed me to use some of their personal notes. Finally, Wyne Schrock's proofreading made this a better book. I thank all of them.

A very special acknowledgment must go to my good friend Jay Stewart for making time in his busy schedule to develop the cost data for the electrical estimating procedure. I could not have done it without his help. Thank you, Jay!

Pablo F. Navarrete

CONTENTS

FOREWORD   **v**
PREFACE   **vii**

1. INTRODUCTION   1

1.1	**Scenario**	1
1.2	**The Small Project**	2
1.3	**Project Execution Overview**	4

Introduction • Initial Involvement/Plan of Action • Process Design (Phase 0/Phase I) • Estimating • Project Execution Plan/Master Project Schedule • Contractor Selection • Detailed Engineering • Procurement • Construction Management • Project Control • Contract Administration • Communications

2. INITIAL INVOLVEMENT AND PLAN OF ACTION   8

2.1	**Initial Involvement**	8

Overview • Memo of Understanding

2.2	**Initial Plan of Action**	9

General • Approximate Construction Hours •Approximate Engineering Hours • Process Design Hours • Project Duration/Peak Staffing • Plan Contents • Planning Rules of Thumb

2.3	**Case Study**	16

3. PROCESS DESIGN - PHASE 0/PHASE I   22

3.1	**Overview**	22
3.2	**Process Design Packages**	22

Conceptual Design • Phase 0 Design • Phase I Design

3.3	**Project Manager's Role**	28

Cost Optimization • Phase I Review • Phase I Specifications

3.4	**Conceptual Plant Layout Guidelines**	31

4. PROJECT EXECUTION PLAN/MASTER SCHEDULE   **36**

4.1   **Overview**   36

4.2   **Thoughts on Scheduling**   38

4.3   **Influential Factors**   38

4.4   **Preparation Guidelines**   39
       General • Preliminary Execution Plan • Case Study • Project Specific
       Durations • Questions/Decisions • Master Project Schedule • Firm
       Execution Plan • Presentation

4.5   **Compressing the Schedule**   50

4.6   **Project Coordination Procedure**   53

5. ESTIMATING   **55**

5.1   **Thoughts on Estimating**   55

5.2   **Estimate Definitions**   58

5.3   **Estimating Methods**   59
       Proportioned Method • Factored Method • Computerized Simulations •
       Detailed Method • Semi-detailed Method

5.4   **Anatomy of an Estimate**   65
       General • Breakdown

5.5   **Cost Allocation**   70

5.6   **Adjustments**   77
       Resolution Allowance • Escalation • Contingency

5.7   **Checking Criteria and Guidelines**   79

6. CONTRACTING   **83**

6.1   **Overview**   83

6.2   **General Considerations**   84

6.3   **Types of Contracts**   84
       By Mode of Selection • By Breadth of Scope • By Mode of
       Reimbursement

6.4   **Contracting Strategy Criteria**   87
       General • Engineering • Construction

6.5   **Selection of EPC Contractors**   89
       Preparation of Bid Package • Bidders Selection • Preparation of Bids •
       Bids Evaluation and Contractor Selection

6.6   **Subcontracting Construction Work**   97
       Overview • Bid Package • Bidders' Qualification • Bidding • Bid Analysis
       and Contract Award

6.7   **Contracting Engineering Services**   101

Contents

6.8	**Do's and Don'ts of Contracting**	103
6.9	**Typical Contracts**	104
	The Agreement • Scope of Work • General Terms and Conditions • Special Terms and Conditions • Proposal Information • Reimbursable Costs Schedule	

7. DETAILED ENGINEERING — **109**

7.1	**Overview**	109
7.2	**Execution by Contractor**	111
	Basic Engineering • Detailed Design •Coordination and Control	
7.3	**Small Project Execution Options**	116
	General Considerations • In-House Engineering • Contracted Engineering	
7.4	**The Project Manager as General Contractor**	119

8. PROCUREMENT — **122**

8.1	**Overview**	122
8.2	**Guideline for Purchasing**	123
8.3	**Expediting and Inspection Criteria**	125
	Expediting • Inspection • Performance Testing	

9. CONSTRUCTION MANAGEMENT — **129**

9.1	**Overview**	129
9.2	**Construction Options**	130
9.3	**Construction Management Activities**	131
	Actual Construction • Construction Management	
9.4	**The Project Manager as Construction Manager**	135
	Overview • Initial C.M. Activities • Recommended Field Reports/Logs	
9.5	**Influence of CICE on Construction Management**	139
9.6	**Co-Employmentship**	140

10. PROJECT CONTROL — **141**

10.1	**Thoughts on Project Control**	141
10.2	**Project Control and the Project Manager**	143
	Cost Control • Schedule Control	
10.3	**Control in the Early Stages**	145
	Site Selection • Phase 0/Phase I • Project Execution Plan/MPS • Contracting	
10.4	**Control in the Engineering Office**	150
	General • Plant Layout • Detailed Engineering • Purchasing and Subcontracting	
10.5	**Control During Construction**	154

10.6	**Control During Project Control**	155
10.7	**Anatomy of a Project Control System**	156
10.8	**In-House Cost Tracking**	158
10.9	**In-House Construction Progress Monitoring System**	159

General • Activity Breakdown • Value System • Schedule • Progress
Computation

10.10	**In-House Engineering Progress Monitoring System**	165

Detailed System • Quick System

10.11	**Cost and Schedule Forecasts**	170
10.12	**Checking Contractor's Schedule and Execution Plan**	171

General • Review Criteria

10.13	**Avoiding/Correcting Frequent Problems**	173

In the Engineering Contractor Office • During Procurement •
During Construction

10.14	**Work Sampling Guidelines**	176

11. CONTRACT ADMINISTRATION   178
11.1	**Overview**	178
11.2	**Thoughts on Contract Administration**	178
11.3	**The Project Manager as Contract Administrator**	179
11.4	**Typical Audit Exceptions**	181

12. COMMUNICATIONS   184
12.1	**Criteria and Guidelines**	184
12.2	**Documentation Checklist**	185

13. SEMI–DETAILED ESTIMATING SYSTEM   188
13.1	**Procedure**	188

General • Order of Magnitude and Conceptual Estimates •
Semi-detailed Estimate

13.2	**Equipment Estimating Procedures**	193

General • Vessels • Pumps • Shell and Tube Heat Exchangers •
Miscellaneous Equipment • Equipment Erection

13.3	**Civil Work Estimating Procedures**	208

Concrete Work • Structural Steel • Miscellaneous Civil Work

13.4	**Piping Estimating**	220

Comprehensive Unit Prices • Miscellaneous Comprehensive
Unit Prices • Miscellaneous Valves Costs

13.5	**Insulation Estimating**	229

Contents

13.6	**Electrical Work Estimating Procedure**	235
13.7	**Instrumentation Estimating Procedure**	241
13.8	**Engineering Hours Estimating System**	245
	Introduction • Hours at Engineering Contractor's Office • Hours to Prepare Phase I Package • Hours to Monitor Contractor's Work • Hours for In-House Engineering	
13.9	**Field Costs**	263
	Labor Costs • Field Indirects • Labor Productivity	
13.10	**Adjustments**	271
	Resolution Allowance Criteria • Escalation • Contingency Determination	
13.11	**Quick Estimate Checks/Conceptual Estimating**	275
	General • Total Installed Cost • Piping Costs • Equipment-Related Costs	
APPENDIX A	Recommended Reading	**280**
APPENDIX B	Glossary	**281**
APPENDIX C	Typical Coordination Procedure	**285**
APPENDIX D	Estimate Checklist	**294**
APPENDIX E	Technical Evaluation Criteria Example	**299**
APPENDIX F	In-House Construction Progress Monitoring System Example	**299**
APPENDIX G	Forecasting Final Subcontract Cost	**308**
APPENDIX H	Heat-Tracing Cost Models	**314**
APPENDIX I	Field Indirects Checklist	**316**
INDEX		*319*

CHAPTER 1
INTRODUCTION

1.1   Scenario

The Owner in the text is assumed to be a multidivision, medium sized chemical company with manufacturing operations in various locations.   The execution of engineering and construction projects is handled by a Corporate Engineering Department (CED) with a large process engineering and control section staffed by a group of project managers, control engineers and a limited number of specialists: estimating, cost control, contracting, mechanical and electrical.   The project managers are usually assigned to the larger projects that are executed in a conventional manner through engineering contractors acting as general contractors. The project engineers are normally assigned either to support the project managers in conventional projects or to act as project managers in smaller projects.

Although CED has a strong process engineering group, it has minimal, detailed design capabilities and relies heavily on engineering contractors and, to some extent, on plant engineering groups, some of which have limited capabilities. CED expertise is in the areas of process engineering and project management and control.

Projects are initiated, developed and sponsored at the Division level.   The Division in turn delegates the execution of engineering and construction to CED, while retaining the overall responsibility for all project activities.   The Division discharges its responsibilities through a venture manager who coordinates the work of the various groups: R&D, marketing, production, CED, etc. In this scenario, the Venture Manager represents the business interests and has the ultimate responsibility for all phases of the project from the marketing studies to the start of the facility.   The physical execution of the project - process design, engineering, procurement and construction - is the responsibility of a CED Project Manager/Engineer reporting to the Venture Manager.   The typical coordination procedure in Appendix C illustrates the responsibility breakdown among the members of a venture team.

On the larger projects, CED takes a conventional approach and assigns an experienced project manager, supported by an in-house project team to direct and supervise the activities of the contractor(s) retained to do the actual work.

CED must also execute many small projects that, if handled in the conventional way, would increase cost and overstrain the limited in-house resources. These projects are assigned to versatile, hands-on engineers (not necessarily experienced project managers) capable of performing personally some of the activities involved in project execution and control.

1.2   The Small Project

The determining criterion to differentiate major projects from small projects is the degree of complexity, not necessarily the project cost. Occasionally, an expenditure of tens of millions of dollars could be simple enough that it might be better handled using the small project approach to conserve spending, rather than following the conventional approach used in major projects.

- A major project involves up to several hundred equipment items and purchase orders versus several dozens in a small project.
- The engineering of a major project requires dozens, maybe hundreds of thousands of home office hours, while a small project rarely requires more than 10-15 thousand.
- Construction hours in a major project usually add up to several hundred thousand. On small projects they are usually less than 50 thousand.

The complexity of the project dictates the execution approach. The coordination of a major project requires a well-structured organization, with formal lines of communication and procedures to insure quality, schedule and cost control. To achieve this, the direct execution and project control responsibilities are assigned to an engineering and/or construction firm to act as general contractor under the supervision of the Owner's Project Manager. In the conventional approach required for major projects, the general contractor is a must.

The activities associated with a small project are basically the same as a major project. However, the reduced complexity does not warrant the elaborate and formal organization provided by a large engineer/construction contractor or construction manager. In small projects, the Owner retains the direct execution and project control responsibilities and uses engineering and construction contractors, as well as in-house resources to perform specific project activities. The Owner becomes, in fact, the general contractor. Substantial savings can be retained through the small project approach when applied to the right size of project.

In any case, the Owner's Project Manager has the overall responsibility for achieving the project objectives (cost, schedule, quality) while keeping management well informed and doing so within the framework of the applicable policies and standards.

In a major project, in addition to the general contractor, the Project Manager is usually assigned a team of engineers specifically dedicated to support the project and follow up on the details of the work to be performed and monitor progress. The Project Manager's role is mainly to provide overall direction, ascertain that proper controls are established and monitor their implementation to insure compliance to project objectives and specifications.

In a small project, there is no general contractor and the staff support will be on a limited as-needed basis. Therefore, in addition to having the overall responsibility, the Project Manager must insure that the vacuum left by the lack of general contractor is adequately filled, and must spend much of the time "in the trenches" doing things or making them happen. Since there is limited support staff to be delegated, the Project Manager must be prepared to take on many roles.

In a major project, the Project Manager is an experienced project manager. whereas the small project manager could be a project engineer or even a process or specialty engineer with a varying degree of project experience. The small project manager must compensate for any lack of experience with a high level of energy and willingness to be immersed in all aspects of the project. The Project Manager for a major project must be:

- A thinker.
- An organizer.
- A delegator.
- Able to maintain perspective.

On the other hand, the small project Project Manager must be:

- Versatile.
- Flexible.
- A doer.
- Someone with a can-do attitude.

Both must be:

- Proactive.
- Decision makers.
- Able to get along with people.
- Very cost conscious!

PROJECT MANAGERS CANNOT ACHIEVE FULL POTENTIAL
WITHOUT A GOOD SENSE FOR COST.

This book will provide all project managers and engineers acting on behalf of
owners effective yet simple tools and guidelines to optimize all phases of project
execution, thus enhancing the overall project performance.

1.3   Project Execution Overview

Introduction

The execution of a project involves a series of activities covering a wide range of
management, engineering and control functions. It is the responsibility of project
managers to insure the thorough and timely execution of those activities in order to
bring projects to a successful conclusion.

To fulfill these responsibilities, project managers must participate actively and
in a deliberate manner in all of them by actually performing some and directing or
coordinating others. This is particularly true in small projects where the Owner's
project team must perform the functions normally done by the general contractor.
The succeeding paragraphs summarize the project activities, discuss the Project
Manager's participation and emphasize the differences between small and major
projects.

Initial Involvement/Plan of Action

Promptly after being assigned to the project, the Project Manager must contact the
project sponsor representative (Venture Manager) to review the project scope and
objectives and determine whether further scoping is required. This meeting should
be documented with a memo confirming the Project Manager's understanding of
the scope and objectives. The Project Manager must then prepare and publish, as
soon as possible, an initial plan of action. This plan of action must address all the
activities required for the preparation and approval of an AFE (authorization for
expenditures) and assign execution responsibilities within the organization. The
plan of action must also address the feasibility of the desired schedule and, when
necessary, sound the alarm and propose remedial alternate solutions.

Process Design (Phase 0/Phase I)

A process design package is the detailed definition of the proposed facility and
must be completed before the detailed engineering activities can start and proceed
effectively. Although the actual process design is done by others, the Project

Manager must participate actively providing project engineering and cost input to insure a cost effective design.

Estimating

Estimates for large projects are normally prepared by staff estimators or a contractor. However, the Project Manager must be capable of at least spot-checking them and ascertaining that sufficient information is provided to prepare realistic execution plans. On small projects, the Project Manager is frequently required to prepare estimates and is expected to be capable of doing so.

The semi-detailed estimating system and related procedures presented in this book will allow project managers to prepare reasonably accurate estimates very quickly and confirm the validity of estimates prepared by others even quicker.

Project Execution Plan/Master Project Schedule

One of the most important project management activities is the preparation of realistic execution plans. Project execution plans can, and should, be prepared for any type of estimate: conceptual, preliminary and definitive. It is the Project Manager's responsibility to prepare them. A thorough execution plan must address:

- Engineering, equipment delivery and construction schedule.
- Interdependence of key activities.
- Contracting strategy.
- Assignment of responsibilities.
- Home office and field staffing (average and peak).
- Base progress curves.

Contractor Selection

Contractor selection is probably the activity that has the most lasting effect on project execution. A poor design or a bad estimate can always be revised and, if caught in time, the effects minimized. Changing a contractor after the work has started is a very difficult proposition that always has a negative impact on project cost and schedule.

The prime responsibility of contracting falls upon the Contract Engineer. However, the Project Manager must live with the selected contractor and make it perform. It behooves the Project Manager to take a very active participation in contracting activities and be thoroughly familiar with all contractual terms and conditions.

Detailed Engineering

Detailed engineering in a small project is a mixed bag and could be executed by a combination of the following:

- Plant engineers.
- Staff engineers.
- Small local engineering firms.
- Large engineering firms.

NOTE: When engineering firms are retained, the scope of their work must be limited to clearly defined design work.

In all cases, the Project Manager must coordinate and monitor the activities of all groups and generally provide the management usually provided by a general contractor.

Procurement

Procurement in a small project is usually a joint effort between CED and the plant's purchasing department. Normally, the Project Manager, with help from CED specialists, as required, writes the requisitions and the plant writes the purchase orders. Expediting and inspections can be provided by either one. However, it is up to the Project Manager to keep up with delivery schedules and take the necessary action to correct slippages.

Construction Management

This activity in small projects is directly performed by the Owner. The Project Manager is also expected to be the Construction Manager. Frequently, the work is delegated to another engineer or to a member of the plant staff, but the overall responsibility remains with the Project Manager.
    The Construction Manager is expected as a minimum to:
- Organize and supervise all field work.
- Coordinate the interface between the various contractors.
- Coordinate construction with plant activities:
    1. Control change orders.
    2. Enforce plant rules and regulations.
    3. Monitor progress.

Project Control

PROJECT CONTROL IS A CONTINUUM AND IS EXERCISED THROUGH THE EFFECTIVE EXECUTION OF ALL PROJECT ACTIVITIES.

The Project Manager is expected to identify and correct cost and schedule variations and make accurate forecasts of both, and to keep management informed on a timely basis. In a major project, the Project Manager's activities are usually limited to monitoring and spot-checking the general contractor's reports. On a small project, the Project Manager must establish and implement control procedures to monitor work progress with minimum effort and reasonable accuracy.

Contract Administration

The Project Manager's concern is not only the physical conduct of the work, but also the implementation of all contractual conditions, especially those asserting the Owner's right of approval and control of the purse strings.

Communications

Keeping management well informed avoids unpleasant surprises and allows it to exercise overall project control. It is essential that the Project Manager report accurately, factually and promptly all problems, errors, significant cost variations and potential problems. The most critical requirement of any report is that it "tells it like it is". Problems and potential problems must be faced realistically to permit timely corrective action.

CHAPTER 2
INITIAL INVOLVEMENT AND PLAN OF ACTION

2.1 Initial Involvement

Overview

Promptly after being assigned to the project, the Project Manager must contact the requester, usually the Venture Manager, to:

- Define and understand the scope - what do you want and when do you want it?
- Determine client needs.
- Gather enough information for the preparation of a sensible initial plan of action.

In a nutshell: "Tell me what you need and I will think it over and be back to you ASAP with my recommendations."

If the assigned Project Manager has limited experience, a senior project manager or a supervisor should also participate in this meeting.

It is very important that this meeting be documented formally in a memo of understanding to:

- Document the basis of the project and its goals.
- Confirm the Project Manager's understanding of the scope and various clients' requirements.

- Keep CED management and appropriate key personnel informed.

The memo of understanding should be discussed with the Venture Manager before issuing it. This will avoid potential arguments and should not take more than a brief conversation.

Memo of Understanding

As mentioned in Preface, this guideline is not intended as a cookbook; it does not recommend the use of standard formats for project memos.

- Each project is unique.
- The circumstances of the first contact with the "client" are different.
- Every Project Manager has a different approach and a different writing style.

As a result, each memo of understanding could be different, but will always:

- Document the scope of work.
- Formalize the initial request of the project sponsor and the expected CED involvement.
- Apprise the Venture Manager of the additional scope information that must be developed before a meaningful planning effort can begin.

2.2 Initial Plan of Action

General

The Project Manager is specifically responsible for planning the overall project execution, ascertaining the validity of the input and establishing realistic cost and schedule objectives. A complete execution plan cannot be finalized until a detailed scope and a good quality estimate have been developed. However, planning must start at project inception so that serious roadblocks can be identified and avoided early in the game.

The Project Manager must develop and document, as a minimum, an initial plan of action addressing the activities required to prepare the funding request and justification, assigning responsibilities, setting a realistic timetable and providing an estimate of the required staff-hours and associated costs required during the early project stages.

The initial plan of action must also address a proposed overall project execution strategy in order to promote creative thinking as early as possible. If the objectives set by the project sponsor are not realistic, it is the responsibility of the Project Manager to bring up the realities of life and, working closely with the Venture Manager, develop achievable goals.

The initial plan of action must be prepared for a project, large or small; and in all cases, it will require the direct hands-on participation of the Project Manager and the review and approval at the appropriate management level.

The scope definition provided during the initial contact may or may not be sufficient to prepare a meaningful plan of action. If not, the Project Manager must work closely with the Venture Manager to develop it.

> PROJECT MANAGERS MUST DO THE BEST THEY CAN WITH THE INFORMATION DEVELOPED IN THE FIRST CONTACT WITH THE VENTURE MANAGER.

The determining parameter for establishing the duration of any project activity is the hours required to execute it. Consequently, the engineering and construction hours, especially the latter, determine the project duration. If they are not available, the project manager must develop them. The scope provided by the project sponsor must contain sufficient information to develop, through literature and/or company files research, the approximate total installed cost (TIC) of the proposed facility and/or equipment count. Both are used in the preparation of conceptual engineering and construction hours estimates.

Approximate Construction Hours

Construction labor costs can be derived from the TIC with the aid of the Lang Factors discussed in Chapter 5.

$$\text{Construction Labor Cost} = \text{TIC} \times 0.38$$

The approximate hours can then be calculated dividing the cost by \$40/hr. which is a typical rate for union labor in industrialized states, including subcontractors, indirects, overhead and profit.

The weakness of using TIC is apparent, but if approximate TIC is the only information available at this time, the Project Manager must make do with it.

A more rational approach can be followed when the approximate equipment count is available. Then:

$$\text{Construction Hours} = \text{Equip. Count} \times \text{Growth Factor} \times 1700$$

The growth factor is a function of the status of the process design:

- Conceptual design   1.30 - 1.50
- Preliminary design (Phase 0) 1.10 - 1.30
- Complete Design (Phase I)   1.05 - 1.10

As discussed in Chapter 13, Section 13.1, equipment count also offers a rational approach to estimate TIC, especially for liquid process organic chemical plants. The scope of equipment count is defined in Section 13.11.

Approximate Engineering Hours

Engineering hours are intimately related to equipment count and type. When a complete equipment list is available, the procedure included in Chapter 13 can be used to prepare accurate estimates. However, this is seldom the case in the early stage of any project and the project managers must do the best they can with the information they have.

If the approximate TIC is available, the engineering hours can be developed with the aid of the Lang Factor.

Engineering Cost = TIC x 0.16
Approximate Hours = Cost ÷ $65/hr.

When the equipment count is available:

Engineering Hours = Equip. Cost x Growth Factor x 650

Process Design Hours

The Engineering hours required for process design are not only related to equipment count and type; they are also very sensitive to the type of process and stage of development of the process design. Experience shows that these factors can produce very wide variations in the hours required.

Since the process design hours determine the cost and duration of the initial project activities, it is very important that the initial plan of action be based on realistic estimates. Unfortunately, as mentioned before, the level of information required to prepare accurate estimates is rarely available at this stage of the project. Nevertheless, this section tries to provide some criteria and guidelines to allow project managers to make the best out of the scant information available.

Analysis of actual cases shows that the hours required to prepare a complete and formally documented process design package (Phase I) including drafting and all required engineering disciplines vary widely depending on the type of project.

Low	Average	High
30 hr/item	75 hr/item	150 hr/item

The following factors determine the location within the range:

High End

- New technology.
- Hazardous service.
- Process complexity.
- Large equipment.
- Low equipment count.
- Process options.
- Execution by contractor.

Low End

- Simple process.
- Simple equipment.
- Repeat project (old Phase I manuals).
- All work in-house.

If equipment count is not available, the approximate hours for a Phase I package can be developed using 12% of the engineering hours derived from the Lang Factor.

The preparation of a preliminary process design package (Phase 0) would require approximately 40% of the hours of a Phase I package.

Project Duration/Peak Staffing

The duration of the construction period as well as the peak staffing required can be determined from Fig. 2.1.

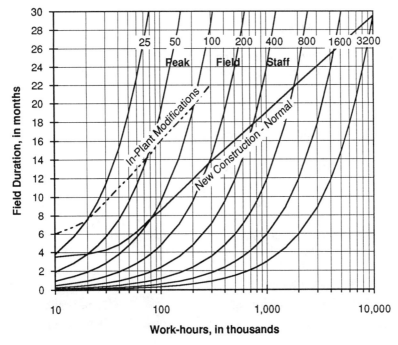

Figure 2.1 Construction duration chart.

Note:
- Field duration from start of pile driving or foundations to mechanical completion.
- Field hours include all direct hired, sub-contractor and supervision hours.

In a typical chemical process plant, the duration of the construction phase is approximately 80% of the total project duration from the start of engineering (after completion of Phase I) to the mechanical completion of the construction work. The remaining 20% is the lead time required for engineering and procurement activities before construction can start in a cost-effective manner.   It must be noted, for initial planning purposes, that the lead time allowed should not be less than three months. Historically, the engineering hours spent during this period are normally between 30% and 40% of the project total.

The total project duration and anticipated engineering and process design staffing can now be easily approximated.

Example

Given:	Construction hours	80,000
	Engineering hours	28,000

From:	Approximate TIC or equipment count	
Then:	Construction duration from Fig. 2.1	7 months
	Peak construction staff	100 workers
So:	Total duration is 7 ÷ 0.8	9 months
And:	Engineering lead time is 9 - 7	2 months
	But 2 months is less than minimum	
So:	Total duration is 7 + 3	10 months
And:	Engineering hours used during lead period are 35% of 28,000	9800 hrs.
Equivalent to:	9800 ÷ 3 mo.	3300 hr./mo.
	3300 ÷ 160 hr/mo.	21 persons avg.
	21 x 1.6	24 persons peak
Also:	Phase I hours 28,000 x 0.12	3400 hrs.
Assuming:	5-6 process engineers can be made available	
Then:	Phase I duration	approx. 4 months

Plan Contents

The initial plan of action must address the following subjects:

- Viability of sponsor's cost and schedule objectives. If required, recommend realistic ones.
- Preparation of the process package: engineering hours, duration and proposed execution method.
- Project funding philosophy and schedule.
- Responsibility for the execution, anticipated CED involvement.
- Project execution approach - small/conventional.

- Approximate engineering and construction staff requirements: average and peak.
- Preliminary schedule.
- Contracting strategy.
- Contractor selection timetable.

Planning Rules of Thumb

The following is a brief summary of the tools, discussed early in this chapter, that will enable project managers to prepare sensible plans of action very early in the project with minimal information:

Construction Hours

- TIC x 0.38 ÷ $40/hr.
- 1,700 hr. per equipment item.

Engineering Hours (After Phase I Design)

- TIC x 0.16 ÷ $65/hr.
- 650 hr. per equipment item.

Equipment Count "Growth"

- Conceptual to final   1.30 - 1.50.
- Phase 0 to final   1.20 - 1.30.
- Phase I to final   1.05 - 1.10.

Process Design (Phase I) Hours

- Engineering hours x 0.12.
- 75 hr. per equipment item.

Total Project Duration

- Construction duration ÷ 0.8.

Engineering Lead Time

- Total project duration x 0.2 or 3 months minimum.

Lead Engineering Hours

- Engineering hours x 0.35.

Peak Staff

- Average staff x 1.6.

2.3   Case Study

Initial Contact

The sponsor asks CED to take over the design and construction of a liquid organic chemical plant, of known technology, to be located at an existing site. Marketing needs dictate that the facility be in production 18 months from the initial contact but management does not wish to spend money on contractors before formal approval of the AFE.

The sponsors R&D Group has prepared a preliminary Phase 0 process design showing 50 pieces of equipment and estimated a total installed cost of around 10 million dollars.

Project Manager's Contribution

TIC Validation

- Anticipated equipment count growth   25%
- Anticipated TIC
  50 items x 1.25 x $224K/item $14.0M
  (from Section 13.11)

The 40% difference should be of concern to the project sponsor but is still within the range of accuracy of conceptual estimating.

Preliminary Construction Hours

- Based on Anticipated Cost:

Low	(10M x .38)/$40 per hr	=	95,000

High	(14M x .38)/$40 per hr	=	133,000

- Based on Equipment Count:
  50 items x 1.25 x 1700 hr/item   =   116,000

Avg: 111,000

Preliminary Engineering Hours

- Based on Anticipated Cost:
  Low   (10M x .16)/$65 per hr   =   24,600

  High   (14M x .16)/$65 per hr   =   34,500

- Based on Equipment Count:
  50 items x 1.25 x 650 hr/item   =   40,600

Avg: 32,200

Engineering/Construction Duration and Peak Staff

Given:	Construction Hours	111,000
	Engineering Hours	32,200
Then:	Construction Duration	9 months
And:	Peak Construction Staff	125
So:	Engineering & Construction Duration 9 ÷ 0.8	11.25
And:	Engineering Lead Time 11.25 - 9	2.25
	But: 2.25 Months is less than minimum	
So:	Engineering & Construction Duration 9+3	12 months
And:	Engineering Hours used during lead period	
	are 35% of 32,200	11,300
Equivalent to: 11,300 ÷ 3 months		3,800 hrs/month
	3,800 ÷ 160 hrs/month	23.5 Avg. Staff
And:	Peak Eng. Staff 23.5 x 1.6	38 Peak Staff

Process Design Hours and Duration

Based on equipment count:	
50 x 1.25 x 75 hrs/item	4,700
Based on engineering hours:	
32,200 x 0.12	3,900

Avg: 4,300

A team of 6 persons would require 4.5 months to prepare the process design.

Total Project Duration

Normal Execution

The completion of this project requires a series of steps that would normally be executed in a consecutive manner.

Step	Duration (Months)	Comments
- Form project team	0.5	Judgment required.
- Select contractor	1.5	To perform Phase I.
- Approve funds for Phase I work	---	$280K to pay for 4300 Hours. Special permission must be obtained to start spending before AFE approval.
- Complete Phase I	4.5	
- Prepare AFE Estimate	0.5	
- Approve AFE	2.0	Dictated by company procedures.
- Start Detailed Eng.	3.0	Minimum lead time.
- Construction	9.0	
- Start up	1.0	Judgment
TOTAL:	22.0	

The total duration far exceeds the 18 months dictated by the marketing requirements. In this particular case study, the picture is complicated by a long delivery item that, under normal circumstances, would be delivered to the field more than 19 months after project start. Refer to Fig. 2.2.

Fast Track Execution

Although the initial pass at the schedule indicates a potential delay of 4 months, the situation can be corrected if management is willing to risk an additional 0.75 million dollars in pre-AFE funding and increase the engineering costs slightly.

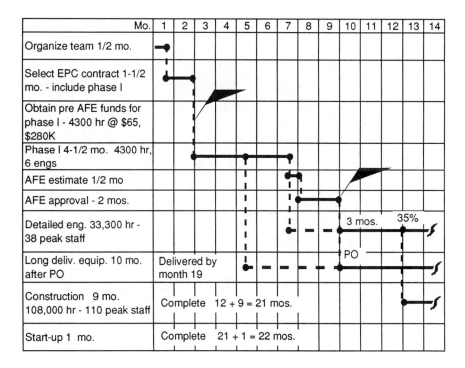

Figure 2.2 Initial plan of action example - normal execution.

The following actions will bring the total duration back to the required 18 months.

Action	Risks
- Reduce Phase I duration to 3 1/2 months by adding 2 process engineers.	Potential loss in productivity would increase the total hours.
- Increase pre-AFE funding to 1.0 million dollars to cover 3 months of detailed engineering and cancellation costs for the long delivery equipment.	Lose all pre-AFE expending if project is not approved after the estimate.
- Start detailed engineering before completing Phase I.	Wasted design hours due to false starts and recycles.

The fast track approach is illustrated in Fig. 2.3.

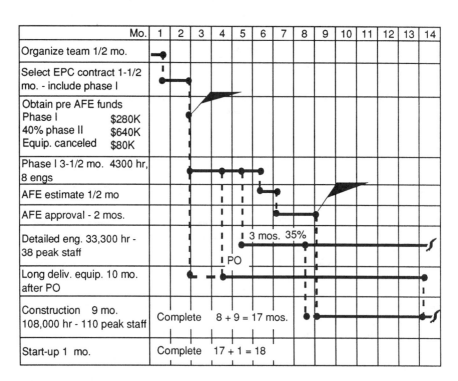

Figure 2.3 Initial plan of action example - fast track execution.

Preliminary Execution Analysis

Execution Approach

The estimated hours for both engineering and construction exceed the limits of the definition of small project in Section 1.2.   This project should definitely be approached as a major project through an EPC contractor.

Size of Engineering Contractor

Ideally, the engineering contractor staff should be 2-5 times the anticipated peak. In this particular case, the bidder list should be restricted to small and medium sized firms with 80 to 160 employees.

Peak Construction Staff

If the site cannot absorb 125 workers and supervision during the peak period, the schedule must be delayed or consideration given to work more than 40 hours per week either by double shift, overtime or some other arrangement.

Construction Subcontractor's Size

The estimated construction hours and the anticipated peak staff suggest that the project could be beyond the reach of the small local contractors.   Consideration must be given to attracting larger contractors or splitting the work and employing several small work contractors.

CHAPTER 3
PROCESS DESIGN - PHASE 0/PHASE I

3.1 Overview

Process design is the first action step in the execution of a project. Process design packages are normally prepared to serve as the starting point for more advanced design stages as well as the basis for the various types of cost estimates required to make progressive project decisions. For example:

- A **Conceptual Design** is the starting point for a basic process design (Phase 0) and can be used to prepare both order of magnitude and conceptual estimates to support initial business planning decisions.
- A **Phase 0 Design** is the starting point for a complete process design and engineering specification (Phase I) and is also the basis for the preliminary cost estimate required to optimize a process and/or decide to continue or cancel it.
- A **Phase I Design** is the starting point of detailed engineering and mechanical design. It is also the basis for the definitive cost estimate required for appropriation and formal project approval.

The determining criteria for falling into any of the three categories of process design is the soundness of the basic process information and/or assumptions, rather than the degree of detail. A process including detailed materials and energy balances and even Piping and Instrumentation Diagrams (P&ID's) will still be a conceptual design if it is based on a theoretical process that has not been proved beyond the laboratory level. On the other extreme, a simple reference to an existing plant of a given capacity can be equivalent to a Phase 0 or even a Phase I if all the conditions are similar.

The actual preparation of the process design is the responsibility of the Owner's process engineering group or a contractor under the direction of a

Technical Manager. However, as mentioned in the Introduction, the Project Manager must participate actively to insure a cost-effective design. A process design cannot be considered complete without project engineering input. The greatest cost saving opportunities are found in the early stages of any project and it is the greatest challenge to the assigned Project Manager to identify those opportunities and influence the process design team to make the most out of them.

During the Phase I stage, the Project Manager is specifically responsible for:

- Planning the overall project execution.
- Setting realistic preliminary schedules.
- Ascertaining the adequacy of the input to estimating.
- Providing on a continuing basis cost and execution viability input to insure optimum project cost and schedule.

In all cases the Project Manager must review, and/or insure review at the appropriate level, all process designs, especially Phase I packages, for compliance to the stated project objectives.

3.2   Process Design Packages

Conceptual Design

A conceptual design:

- Permits the process engineering group to start developing the technical data and perform the design work required for a basic process design (Phase 0).
- Allows a cost engineer to put an approximate price tag on the proposed facility.
- Provides the business people with sufficient information to assess the risks involved and make meaningful economic evaluations.

A conceptual design could be based either on the application of proven technology and unit operations or, more frequently, on new and incompletely demonstrated technology.

In the first case, the required detail of information is minimal and a decent order of magnitude estimate can easily be developed through a literature search. Preparing any kind of estimate for an unproven process is a different story. A process configuration must be conceptualized so that equipment lists, including estimated sizes and construction materials, can be developed to serve as the basis for an estimate.

The following is a suggested outline of the information that should be included in conceptual design packages for unproven processes:

General

- Description of the proposed facility.
- Purpose for which it is intended.
- Function or products.
- Desired completion date.
- Size or capacity.
- Proposed or possible location.
- Alternative cases which are being considered.
- Mention of any previous scopes, designs, estimates or actual costs for the same general type of project. Differences between previous project and this one.
- Basic philosophy of project: e.g., temporary, long-term, fast track or minimum cost.
- Other facilities or operations at proposed site.

Technical

- Major raw materials, products, by-products, wastes and effluents (compositions or specifications insofar as known).
- Annual design capacity for each product or grade.
- Approximate usages and efficiencies (raw materials and energy).
- Brief process description.
- Health and safety information.
- Environmental considerations.
- Process configuration (preliminary process flow diagram).
- Preliminary process equipment list with approximately sizes and best guess of construction materials.
- Use of existing facilities or services.
- Off-site requirements, including, e.g., storage, handling, shipping and utility supplies.

- Status of process definition.
  1. Unproven/undemonstrated chemical reaction.
  2. Laboratory piloting.
  3. Large scale piloting.

Phase 0 Design

A Phase 0 design provides the general definition of a proposed chemical processing facility. It provides the basis for development of operating costs and economics for business decisions. A Phase 0 design is the starting point for development of detailed process design and engineering specifications (Phase I) and is the basis for a preliminary cost estimate.

The Phase 0 design is normally based on information available from research and development work, experience with a commercial unit, or from an outside licensor. In the latter two cases, there might be abundant knowledge and experience, but for significantly different capacity, local conditions or other factors which require an essentially new design. The following criteria give general and state-of-the-art information which would permit preparation of a Phase 0 design:

- The chemistry has been demonstrated. Major reactions are known. There is some knowledge of side reactions and important by-products.
- A basic flow diagram has been drawn.
- Yields are known, at least approximately.
- Separations and purifications, if demonstrated at all, may have been done only by laboratory techniques which are not necessarily realistic for commercial practice.
- Principal process steps and major equipment have been described.
- Usually, though not necessarily, the most economical materials of construction have been identified.

A Phase 0 package should include essentially the same items suggested for a conceptual design package. However, the information is no longer the result of conceptual work, but is based on hard data derived from actual pilot and/or commercial operations:

- The process configuration is based on complete process flow diagrams (PFD's) and material balances that show the main recycle streams and control loops.
- The equipment list includes all the major process equipment sized to reflect the material balances.

- The equipment list includes basic descriptions adequate for budget pricing.
- The PFD's and material balances identify all waste streams and, as a minimum, propose viable disposal methods and/or schemes.
- The PFD's and material balances also identify the required utilities and provide approximate consumptions.
- The instrumentation philosophy and type must be specified.
- All the required buildings must be listed with brief descriptions and preliminary sizes.
- The preliminary layouts should include all equipment and buildings, show paved areas and roads, and identify on-site requirements.
- Incomplete and/or soft areas subject to significant change must be clearly identified.

Phase I Design

The Phase I design is the detailed definition of the project. Ideally, a Phase I package should contain sufficient information to:

- Have a competent contractor design and build the proposed facility without further input from the client.
- Permit a cost engineer to prepare a preliminary or even appropriation-type cost estimate.

The Phase I design is normally based on a Phase 0 package and on a fully demonstrated process, including waste and effluent handling. The process may have been demonstrated either in actual plant operation or through a sufficiently large pilot operation.

At the end of Phase I, the project must be completely defined:

- Plant location and size as well as the expandability philosophy have been set.
- Raw materials sources have been identified and tested.
- Process steps have been optimized for material and energy efficiencies.
- Recycles and purges have been realistically tested or simulated for control.
- Instrumentation and control systems as well as degree of computerization have been established based on preliminary instrument engineering.

- Product quality and grades have been established; sample or market-test batches have been prepared.
- Product specifications and test methods have been set.
- Packaging and shipping methods have been specified or selected.
- Physical and chemical property data are available for all materials to be handled.
- Materials of construction have been selected or narrowed to an acceptable number of alternatives.
- Energy saving options have been studied and selections made.
- Any remaining process uncertainties have been analyzed and defined.
- Environmental compliance requirements have been established.
- Appropriate hazard analyses have been made, and the impact on design reported. Special hazard protection devices have been defined.
- Insurance requirements have been determined.
- Process engineering including material and energy balances as well as equipment sizing have been completed.

In addition to the information listed under Phase 0, the Phase I package must include:

- Complete physical and chemical data of the raw materials, products, by-products, intermediate products and wastes.
- Complete material and energy balances including average and peak uses of all utilities.
- Capacity turn-down requirements.
- Projected future capacity and philosophy of expandability.
- Complete piping and instrumentation diagrams (P&ID's) showing all equipment, instruments and sized process and utility lines; all properly numbered.
- Preliminary equipment arrangement drawings and proposed layout of required buildings.
- Design data sheets for all equipment items.
- Complete list of instruments with basic description and, when appropriate, manufacturer and catalog number.
- Complete motor list showing estimated total connected and average loads.

- Equipment and piping insulation requirements that can be shown on the P&ID's.
- General specifications for:
  1. Equipment.
  2. Piping and valves.
  3. Electrical.
  4. Insulation.
  5. Other.

3.3   Project Manager's Role

Cost Optimization

During the process design stage the Project Manager can start controlling the final project cost and avoid unpleasant surprises by:

- Providing on-the-spot assessment of the cost and schedule impact of the various process alternatives considered during process design.
- Tracking cost and making projections as the design progresses, thus providing sound criteria for timely action by the Technical Manager and, when required, Owner's management.
- Insuring that all factors with potential cost impact are taken into account so that they can be included in the cost estimates and considered in the project schedule.
- Helping the Technical Manager in the development of cost-effective layouts and equipment arrangements.
- Insuring that the specifications are adequate and do not include unnecessary gold plating.
  1. Excessive corrosion allowances and minimum vessel thickness.
  2. Ultraconservative materials of construction.
  3. Superfluous blocks and bypasses for control valves.
  4. Unnecessary installed spare pumps when process allows for short unit shutdowns.
  5. Structural design for concurrent adverse conditions.

Phase I Review

The Project Manager is not supposed to be an expert in every project-related field but should be knowledgeable and astute enough to ask the right questions, promote constructive thinking and ascertain if others have done their homework. The Project Manager should have a healthy skepticism, be bold enough to challenge the experts to justify their position and/or recommendations, and be adamant in demanding quality work from team members as well as contractors.

The following is not intended as a complete checklist but a reminder for the Project Manager of the important points to be covered as part of a Phase I package review.

- Compliance to the process design and engineering specification.
- Input/review by appropriate groups/individuals:
    1. R&D.
    2. Production.
    3. Maintenance.
    4. Environmental.
    5. Corporate Health/Safety/Hazards.
    6. CED Specialists.
- Schedule viability.
- Adequate information to estimating.
- Coordination with plant:
    1. Work in operating areas.
    2. Shutdown.
    3. Tie-ins.
- Constructability.
- Reasonable specifications.
- Challenges:
    1. Minimum vessel thickness.
    2. Need for high price alloys.
    3. Installed spares.
    4. Expanded layouts.
    5. Winterizing practices.
- Soft areas/allowances.

Phase I Specifications

The specifications define the quality of the design and fabrication of the equipment in the project as well as the related commodities - piping, civil, electrical, etc.   It would be impossible to build a plant without some sort of specifications.

Specifications are also a very important factor in the capital cost of a project since they define design criteria, materials of construction, design safety factors, etc., all of which can influence cost substantially.   The Project Manager controls both quality and cost by ascertaining that the specifications included in Phase I design reflect the quality level established in the project scope as well as the environmental and safety standards mandated by company management and/or applicable laws and regulations while avoiding extravagant and costly requirements.

Specifications can be classified in two general groups:

General Specifications/Standards

These define design criteria, applicable codes and preferences consistent with code requirements.   They should reflect the minimum quality requirements while setting a limit to costly practices.

Project Specifications

These must be prepared for each project to reflect its specific requirements - process, site conditions, safety, specific equipment, etc.   A complete project specification must go into the details of equipment design datasheets, spelling out all the information, process and/or mechanical, required to fabricate a piece of equipment.

Ideally, a Phase I package should include a complete Project Specification. However, this is not always possible.   Some owners don't have the in-house expertise in all the design disciplines to cover the necessary details.   Normally the specifications included in the Phase I packages concentrate on process-related requirements and certain critical mechanical items, leaving most of the mechanical and structural details to the engineering contractor.   On many occasions, specifications from previous projects are included in Phase I packages.   This is a dangerous practice since they are not necessarily updated and/or applicable to the current project.   The Project Manager must avoid this trap carefully and always keep in mind that a good general specification is much better than an outdated, or even incorrect, detailed specification.

On large projects, where detailed engineering is performed by an outside firm, the Project Manager can rely on the engineering contractor to critique, complete and/or update the Owner's specifications as required.   On small projects, the Project Manager is at a great disadvantage and must compensate by being continuously cost conscious and bold enough to challenge the "experts".

3.4   Conceptual Plant Layout Guidelines

General Considerations

Preliminary plot plans and conceptual equipment layouts must be developed during the Phase 0/Phase I work in order to determine the plot requirements and help the estimators visualize the proposed facilities and prepare realistic cost estimates.
This information will be useful for estimating:

- Site development - roads, railroads, fences, etc.
- Civil work - foundations, buildings, structures, etc.
- Tank farms and other off-site facilities.
- Process and yard piping and other distribution systems.

Preparation of final, optimized layouts is a time-consuming process requiring participation of many experienced people. On the other hand, conceptual layout adequate to meet the above objectives can be easily prepared by a process or project engineer. The ensuing guidelines will be useful for doing so. They should also be useful for reviewing layout work done by the contractor.

The recommended parameters and criteria included in these guidelines reflect insurance and OSHA requirements as well as good practice standards. They are biased on the conservative side and their application should result in a layout that could be optimized during detailed engineering without unpleasant surprises.

The following information is the minimum required for the preparation of a conceptual layout. It is occasionally provided as part of a conceptual design package and must certainly be provided in the Phase 0 package:

- Equipment list with approximate dimensions and motor horsepower.
- Process flowsheets or preliminary P&ID's showing relative elevations.
- Off-site requirements - buildings, tank farms, diked areas, railroads, cooling towers, storage areas, etc.
- Hazard considerations.
- Process buildings and/or structure requirements; open/closed.
- Future expansion considerations.

Safety Considerations

Safety and protection of human lives must be the paramount concern in the preparation of layouts. The following considerations are by no means totally inclusive; they are only those that would have a substantial impact on the plant dimensions and capital cost:

- For adequate fire protection, any part of the plant should be accessible from at least two directions.
- Process units handling flammables should be separated to minimize possible spread of fires.
- All areas protected with sprinkler or deluge systems must be contained with provisions for rapid drainage to large remote reservoirs.
- Process vessels with substantial inventories of flammable liquids must be located at grade.
- Flammable materials must be stored outside in diked areas and away from process units and other active areas, control room office change rooms, etc.
- Equipment subject to explosion hazard must be set away from occupied buildings and areas.
- All operating areas must have at least two means of access and exit. On elevated areas, at least one should be a stairway.

Maintenance Considerations

A plant that can not be maintained properly will deteriorate gradually and eventually become inoperable. Good plant maintenance ability is the result of well-thought-out layouts. The following guidelines affect plant dimensions and must be kept in mind when developing layouts:

- All equipment should be accessible by either crane or lift truck.
- Space for maintenance and dismantling must be provided around compressors.
- Bundle removal space must be allowed for shell-and-tube exchangers.
- Sufficient suction head must be allowed for pumps handling hot liquids.

Recommended Minimum Clearances

General

- *Property Line:* All process units and auxiliary buildings, except the front office and the guard house, must be at least 30 ft. from the property line.

- *Primary Roads:* Width 30 ft.; headroom 22 ft.; distance from buildings and process areas 10 ft.
- *Secondary Roads:* Width 20 ft.; headroom 20 ft.; distance from buildings and process areas 5 ft.
- *Pump Access Aisle Ways:* Width 12 ft.; headroom 12 ft.
- *Process Areas Main Walkways:* Width 10 ft.; headroom 8 ft.
- *Process Areas Service Walkways:* Width 4 ft.; headroom 7 ft.
- *Stairs:* Width 3 ft.
- *Railroads:* Headroom 24 ft.; clearance from tract centerline to obstructions 10 ft.
- *Main Pipe Racks:* Headroom 22 ft.
- *Secondary Pipe Racks:* Headroom 15 ft.
- *Floor Elevations:* The vertical distance between operating levels must be no less than the height of the tallest process vessel plus 8 ft. or the tallest tank plus 6 ft.

Around Hazardous Areas

- Flare stacks	100 ft.
- Cooling towers	100 ft.
- Medium flammability liquid storage	50 ft.
- High flammability liquid storage	100 ft.
- Explosion potential	100 ft.

Around Process Equipment

- *Tank Farms:*

1. Between tanks	1/2 dia.
2. From tank to dike wall	5 ft.
3. Access around diked area	10 ft.
4. Dike capacity	Largest tank plus 10%

- *Around Compressors.*   10 ft.
- *Between Adjacent Vertical Vessels:*

1. 3 ft. dia.	4 ft.
2. 3-6 ft. dia.	6 ft.
3. Over 6 ft. dia.	10 ft.

- *Between Adjacent Horizontal Vessels:*
  1. Up to 10 ft. dia.   4 ft.
  2. More than 10 ft. dia.   8 ft.
- *Between Horizontal Heat Exchangers.*   4 ft.
- *Between Vertical Heat Exchangers.*   2 ft.
- *Around Fired Heaters.*   50 ft.

Miscellaneous Equipment Dimensions

- *Pumps*:
  1. Up to 3 HP   1 ft. x 3 ft.
  2. 10 HP   1.5 ft. x 4 ft.
  3. 30 HP   2 ft. x 5 ft.
  4. 75 HP   2 ft. x 6 ft.
  5. 200 HP   3 ft. x 7 ft.
- *Compressors*:
  1. Up to 50 HP   3 ft. x 6 ft.
  2. 100 HP   4 ft x 8 ft.
  3. 250 HP   6 ft x 12 ft.
  4. 500 HP   6 ft. x 16 ft.
  5. 1000 HP   6 ft. x 20 ft.
- *Heat Exchangers*: use nomograph in Fig. 3.1.

Typical Building Dimensions

- Control room   20 ft. x 40 ft.
- MCC room   20 ft. x 40 ft.
- Transformer switch gear   20 ft. x 30 ft.
- Maintenance shop   30 ft. x 60 ft.
- Gate house/first aid   30 ft. x 40 ft.
- Male/female change rooms   30 ft.^{2}/employee/shift
- Lunch room   20 ft.^{2}/employee/shift
- Office building   300 ft.^{2}/employee

Figure 3.1 is also used to approximate the external area in order to estimate the cost of insulation when applicable.

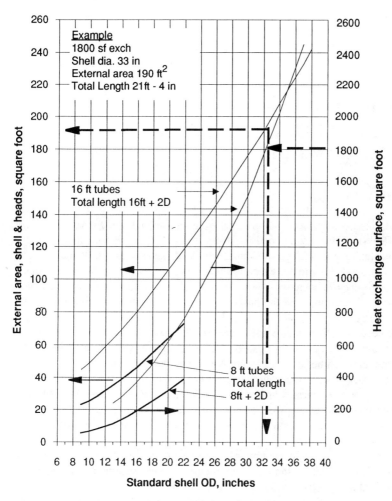

Figure 3.1 Shell diameter & external area.

CHAPTER 4
PROJECT EXECUTION PLAN/
MASTER SCHEDULE

4.1   Overview

Trying to execute a project without an execution plan and a master project schedule (MPS) is like trying to drive from here to there without a map and a clock. Eventually you will get nowhere, you will be late getting there and you won't even know it. A project execution plan combined with the MPS provides a road map with criteria to judge where you are at any time and know when you have arrived. It also provides the basis for the control system and is an excellent communication tool.

The project execution plan is intimately related to the cost estimate; each one depends on the other and neither can be totally complete without the other. For example, the total number of construction hours may be a determining factor in the contracting strategy while the contracting strategy will affect productivity, which in turn impacts on the hours. Furthermore, both will influence schedules. Ideally, each type of estimate, conceptual, preliminary, definitive and/or detailed, should be accompanied by a corresponding execution plan and schedule to complement the scope of work.

The level of a project execution plan discussed in this book is basically a master bar chart schedule showing the required labor loading, indicating the contracting strategy and the expected rate of progress.

Each project has different requirements and unique problems which dictate different approaches. However, all execution plans must basically address and convey the following information:

Preliminary Activities Schedule:

-   Process design.

- AFE estimate and approval.
- Engineering contractor selection.

Procurement Schedule:

- Purchase and delivery of equipment emphasizing long delivery items.

Engineering Schedule:

- Overall summary of engineering broken down by principal design activities.

Subcontracting Strategy and Schedule:

- Most important subcontracts indicating type of contract (lump sum/ reimbursable, unit price, competitive/negotiated).

Loaded Construction Schedule:

- Overall summary of construction broken down by conventional field activities showing the spread of the estimated hours over the duration of each.

Total Staffing Curve.

Construction Progress Curve.

Conceptual and preliminary execution plans can certainly be covered in 20 or 22 activity lines and, therefore, can be documented on one page. Definitive (sometimes) and detailed (always) plans could require several dozen activity lines and several pages. However, for the benefit of management, they must be summarized on one page. The Project Manager must use ingenuity to come up with a meaningful summary.

Project execution planning is one of the most important responsibilities of the Project Manager in both large and small projects. In the early stages of all projects, it requires hands-on participation. In large projects, the execution responsibility is eventually assigned to a general contractor who will be responsible for engineering and/or construction. Detailed planning will then be done by the contractor under the supervision of the Project Manager. The managers of small projects don't have the luxury of a general contractor and they have to cope with all levels of planning. The criteria, tools and guidelines included in this book can be useful to all project managers but are specifically intended for the managers of small projects.

> THE PROJECT MANAGER MUST KEEP A VERY OPEN MIND CONCERNING THE PROJECT EXECUTION APPROACH AND NEVER FORGET THAT NO TWO PROJECTS ARE EXACTLY ALIKE. THE FACT THAT A GIVEN APPROACH WAS SUCCESSFUL IN ONE PROJECT IS NOT, PER SE, SUFFICIENT JUSTIFICATION TO USE IT IN ANOTHER.

4.2   Thoughts on Scheduling

It seems that once a project is initiated, everybody wants to see a schedule immediately. It is up to the Project Manager to provide one. There are all types and levels of schedules ranging from simple bar charts and/or logic diagrams to very complicated computerized networks. All of them have a niche in project execution.

Usually the Owner's project managers don't get directly involved in complex and sophisticated schedules; when the size of the project requires that level of scheduling, there will be an engineering firm and/or a general contractor to perform the required work. The level of scheduling handled by the Owner's project managers, especially those in charge of small projects, should not go beyond that required for a master schedule involving 100-200 activities at the most. That is all that is required to develop a project execution plan.

It must be noted that while a schedule is a stand alone document, the execution plan could not exist without the master project schedule.

4.3   Influential Factors

All the factors listed below can affect project cost and/or schedule. Some will have a favorable impact, others a negative one. Some are controllable, while others are not. Since no two projects are alike, the effect will vary with the circumstances surrounding each one. Every factor must be considered and analyzed before finalizing the project execution plan. The Project Manager must "make things happen", make the most out of the favorable and controllable factors while striving for ways to minimize the impact of the negative ones and working around the uncontrollables.

- Process design execution.
- Execution approach - small/conventional.
- Contracting/subcontracting strategy.
- Schedule - normal/fast track.

- Permits - environmental/construction.
- Procurement:
    1. Long delivery items.
    2. Expediting.
    3. Shipping.
- Engineering hours:
    1. "Normal" duration.
    2. Average/peak staffing.
- Construction hours:
    1. "Normal" duration.
    2. Average/peak staffing.
    3. Availability of qualified workers.
- Modular construction.
- Weather conditions.
- Local economy.
- Labor source - union/open shop.
- Site conditions:
    1. Work in operating areas.
    2. Shutdown work.
    3. Laydown/parking areas.
    4. Welding permits.
    5. Working hours.
    6. Local contractors.

4.4   Preparation Guidelines

General

As mentioned before, the execution plan and master schedule go hand in hand with the cost estimate. Neither can be completed without the other. The project execution plan is the best tool to identify and avoid pitfalls early in the project. It must be prepared as early as possible. A preliminary plan, based on whatever available information, will do the job until a firm plan can be prepared.

If one worker requires ten days to perform one activity, it may be argued that ten workers could perform it in one day. Unfortunately, that is not the case in project execution; all projects involve hundreds, even thousands, of activities, many of which must be performed sequentially. If the activities to be performed in ten days were ten sequential ones instead of a single activity, ten workers could not perform them in one day. More likely, they would perform them in three or four days expending thirty or forty man-days instead of ten. The shortest possible schedule is dictated by the longest string of sequential activities as well as the minimum chronological time and labor hours required to perform each one. The duration of the construction effort is dependent on a number of factors:

- Total construction hours.
- Availability of:
    1. Accurate design and installation details.
    2. Necessary equipment and materials.
    3. Construction equipment and tools.
    4. Qualified personnel.
- Size of construction area.
- Weather conditions.
- Number of hours worked per week.

Fig. 2.1 included in Chapter 2 represents the average of many chemical and pharmaceutical projects. It relates the construction hours to construction duration for "normal" grass roots and retrofit projects indicating the peak staffing expected. It assumes the timely availability of accurate design information, materials and qualified construction personnel. This chart, together with the planning rules of thumb included in Chapter 2, is used as the starting point for all execution plans.

Preliminary Execution Plan

A firm Execution Plan cannot be firmed up until:

- The appropriation estimate and the master project schedule are available.
- The execution approach - small or conventional - has been decided.
- Long equipment and materials deliveries have been determined.
- The engineering and construction contracting strategies have been established.
- The timing of environmental and construction permits has been firmed up with the pertinent agencies.

However, the preparation of execution plans is a bootstrap operation where a preliminary plan must first be developed to make the execution decisions which in turn may impact the original data and determine the firm execution plan. The information required to prepare a preliminary project execution plan is essentially the same as that required for the initial plan of action discussed in Chapter 2 except that it is now based on more accurate data. Whereas the later is based on conceptual data and educated guesses, the preliminary execution plan must reflect, at least, a Phase 0 design and a preliminary or conceptual cost estimate.

A semidetailed estimate prepared by the Project Manager as indicated in Chapter 13 would provide the engineering hours, as well as the total direct construction hours broken down by major construction activities (site work, concrete, structural steel, etc.).

It must be noted at this point that Fig. 2.1 is based on total field hours, direct and indirect, and that the construction hours shown in the estimates must be adjusted before being used to determine project durations.

The direct hours must be adjusted by adding 15% to account for supervision and miscellaneous craft labor (refer to Section 13.9) plus the contingency included in the estimate.

Project durations and peak engineering staffing can now be estimated with the aid of Fig. 2.1 and the pertinent rules of thumb in Section 2.1. Preliminary construction staffing and progress curves can be developed following the guidelines in Fig. 4.1 as illustrated in the following section.

Case Study

The project initiated in the case study in Chapter 2 has proceeded through a Phase 0 process design and the preliminary estimate illustrated in Table 5.4. Now the Project Manager must check the validity of the conceptual data developed for the initial plan of action and prepare a preliminary plan of action for the entire project.

Validity Check

	Conceptual	Preliminary Estimate
- TIC $ M	14.0	13.8
- Field hours	111,000	111,600(1)
- Engineering hours	32,200	36,000(2)

(1) Direct Hours		77,500
	15% Supervision	11,600
	5% Start up Assist.	3,900
		93,000
	20% Contingency	18,600
		111,600

(2) 30,000 x 1.20 = 36,000

EXCELLENT CHECK!

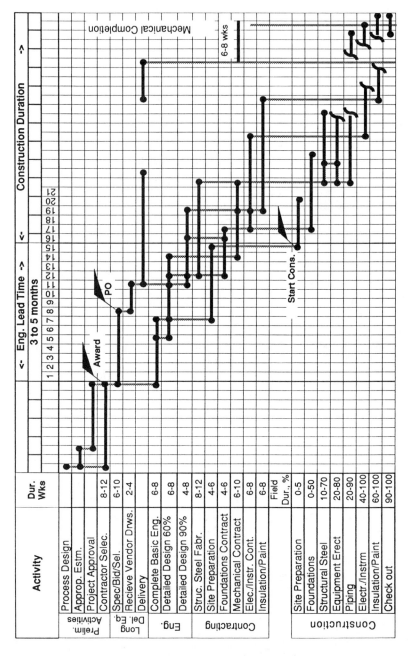

Figure 4.1 Scheduling guideline.

Preliminary Execution Plan

Since the construction and engineering hours have not changed, there is no need to recalculate the durations and the conceptual schedule submitted with the initial plan of action.

However, the breakdown of the construction hours by disciplines provided with the preliminary estimate enables the Project Manager to refine, with the aid of the scheduling guidelines in Fig. 4.1, the staffing requirements and prepare a firm plan for contracting the construction work.

This can be done easily by first laying out the duration of the various construction activities over the estimated 9 month construction duration period in the sequence indicated in Fig. 4.1. The estimated hours for each activity are then distributed over their corresponding durations and totaled, as illustrated in Fig. 4.2, to draw the projected construction staffing and progress curves. The staffing curve peak is 103 versus the 125 estimated with the rules of thumb in Section 2.2.

The preliminary execution plan must then be analyzed to determine its compatibility with the realities of local weather conditions, plant production schedules and other constraints specific to the project.

Project Specific Durations

The durations developed so far represent an average for "normal" project execution and must be adjusted to reflect specific conditions, mostly weather and plant operations.

Weather

- Excavation and concrete work in freezing weather can be very expensive and should be avoided. In some instances the construction site may have to be shut down during the winter months.
- Outdoor mechanical work in rainy and/or extremely cold weather could be a safety hazard and should be restricted unless temporary sheltering is provided.

Operating Restrictions

- When plant shutdowns are involved construction durations are mostly dependent on the production schedules.

- Construction work in active operating areas, if at all allowed, usually must be performed with limited crews at a low productivity level and its duration will extend well beyond "normal".

After the durations have been adjusted, the estimated hours adjusted for productivity, if required, are distributed over the new durations to develop a revised peak staffing and construction progress curve.

Questions/Decisions

The revised preliminary execution plan promotes thinking and raises questions that must be answered in order to refine the plan and prepare an accurate appropriation cost estimate.

- Can the plant facilities absorb the peak construction and supervisory staff without impairing the safety of the production activities? Can so many workers perform efficiently in the designated construction areas? If not, extended overtime or shift work must be considered. Otherwise the schedule must be extended.

    NOTE: Many contractors consider that a labor density of 5 workers per thousand sq. ft. based on total construction area is a reasonable allowance for good productivity.

- Could a single local contractor handle all the work efficiently? If not, consider national general contractors and/or break construction into discrete portions that could be handled by the available local contractors.
- Can the Owner come up with sufficient qualified personnel to perform the Phase I design in a reasonable time - say 3 or 4 months? If not, Phase I must be contracted or extended to match the available resources.
- Since the engineering hours are well beyond the range of the small project approach, the execution plan should be based on retaining a full service engineering firm with capabilities. Based on the anticipated peak engineering staff, the contractor screening should be limited to firms with staffs ranging from 150 to 400 (2 to 5 times the expected peak).

Master Project Schedule

The master project schedule (MPS), like the appropriation estimate, is a stand alone document that must be completed before a firm execution plan can be developed.

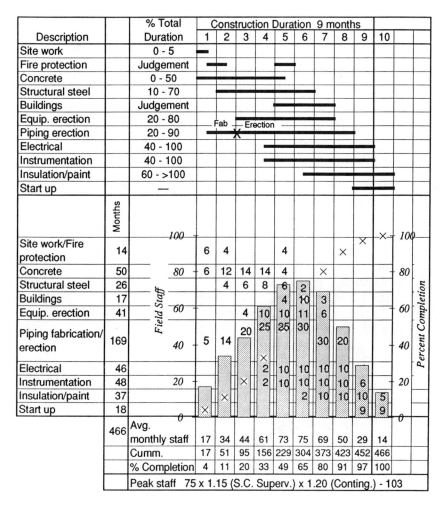

Figure 4.2 Preliminary execution plan progress curve/field staff.

This section regarding the preparation of the MPS offers very simple, yet effective, guidelines for its development and proposes a level of details that would satisfy the Owner's needs in most projects.

Fig 4.1 summarizes the guidelines for MPS development. It is essentially a typical bar chart schedule showing the principal project activities, their interdependence and their normal range of durations. Small projects will fall on the low end of the ranges, while very large projects could very well exceed the high end. The guidelines for field activities are of a very general nature and must be specifically analyzed for each individual project.

The master schedule should include the following general activities. Further breakdown is not warranted at this time.

Preliminary Activities

- Process design.
- Appropriation estimate.
- Funding approval.
- Engineering contractor selection: bid package, bidding, evaluation and award.

Procurement Activities (All Process Areas)

- Specify/bid/P.O./vendor dwg/delivery:
  1. Vessels, as a group.
  2. Pumps, as a group.
  3. Heat exchangers, as a group.
  4. Special and long delivery equipment, individually.

Engineering Activities (for Each Process and Utility Area)

- Basic engineering:
  1. Approved for design P&ID's.
  2. Arrangement drawings.
- Civil design - 60% and 90% complete.
- Mechanical design - 60% and 90% complete.
- Electrical/Instrument design - 60% and 90% complete.

Subcontracting Activities (for Each Subcontract)

- Bid package, bid, evaluate and award.

Construction Activities (for Each Process and Utility Area)

- Site preparation.
- Foundations.

- Structural steel.

- Equipment erection.

- Piping - field fabrication, erection, heat tracing, testing.

- Electrical - power wiring, lighting, check out.

- Instrumentation - installation, loop checking.

- Insulation.

- Painting.

- Final check out (by Owner).

A master schedule prepared as indicated above would include 70-85 activities for a single area project and 170-180 activities for a project divided in three process and utility areas. Those activities could, respectively, be condensed into 30-35 and 65-70 lines of activities.

Firm Execution Plan

Once the basic execution decisions have been reached, the long deliveries determined, the MPS prepared and the appropriation estimate completed, the firm execution plan can be prepared. The plan must include sufficient execution details and schedule information to become the baseline for an in-house progress monitoring system. Such a plan can be prepared with a reasonable effort from the information contained in a Phase I design package and a semi-detailed cost estimate.

The construction activities are broken down to reflect the MPS and the appropriation estimate breakdown and "loaded" with the corresponding hours from the estimate. The resulting construction progress curve becomes the base for the in-house monitoring system discussed in Chapter 10.

The minimum capital cost execution plan is rarely consistent with the optimum cost when marketing requirements are taken into consideration. Sometimes a delay in the completion date may result in loss of several million dollars in sales if a campaign is missed. In this case, the project driving force is a push for early completion, even at the expense of higher capital costs, and the execution plan must be tailored to incorporate schedule reduction steps. The firm execution plan must reflect the most cost-effective approach consistent with the schedule demands imposed by marketing and other business considerations.

If the project duration were longer than the project requirements and the site can absorb a higher manpower loading, the recommended approach would be to find ways to increase the labor density without productivity loss.

Increasing labor density by overloading the work areas without impacting the productivity can only be done to a certain extent. After work areas reach the saturation point, productivity falls dramatically to the point where an increase in forces will result in a decrease in the actual work output. The most practical way to increase labor density without affecting productivity is to open up new work areas and assign additional crews of cost-effective size.

If the cost of execution were the only consideration, the optimum schedule would also be the minimum cost case, and any schedule compression would automatically result in a cost increase. When that is not the case, the obvious explanation is that the initial schedule was not the optimum one. Normally, any schedule is susceptible to compression, at a reasonable cost, up to a point. When that point is reached the cost of further compression results in exponential cost increases and very soon additional improvements become a physical impossibility, regardless of how much money and other resources are applied to the project.

Trying to compress the schedule by more than 10% could be very dangerous. Studies on the subject suggest that while a 10% reduction would impact the project cost by 3-5%, a 20% reduction could be unaffordable for most projects.

The Project Manager developing the project execution plan has the responsibility to:

- Determine the minimum cost schedule.
- Evaluate the various schedule/cost trade-offs to develop a reasonable risk compressed schedule and estimate the associated cost penalty.
- Make sure that management is aware of and understands the risks and the magnitude of the additional costs.
- Incorporate the costs of compressing the schedule into the estimate and adjust the contingency to reflect the related risks.

Presentation

A detailed project execution plan prepared by a contractor for a large project can take dozens, maybe hundreds, of pages breaking the schedule and the work into minute details. This level of effort may be cost-effective for certain critical projects, but could be wasteful for many others. After all, it is not easy to foresee the future in precise details. Fortunately for the Owners' project managers, they don't have to go into that level of detail. However, their execution plans must be concise, yet clear, and convey all the essential information outlined in Section 4.1.

A firm execution plan for a medium sized chemical facility with up to three process and utility areas should not require more than six pages:

- A four page bar chart MPS, including 150-200 activities condensed in 60-80 lines.

- A single page memo highlighting the selected contracting strategy and any out of ordinary schedule maintaining scheme.
- A single page bar chart summary.

The summary must show:

Preliminary Activities

- Process design, AFE estimate, AFE approval.
- Engineering contractor selection - bid package, bid, evaluation and award.

Procurement

- Long delivery equipment - specifications, bid, P.O., vendor drawings and delivery.

Engineering (All Areas)

- Basic design - approved P&ID's, plot plans and equipment arrangements.
- Civil design at 60% completion and 90% completion.
- Piping design at 60% completion and 90% completion.
- Instrument and electrical design at 60% completion and 90% completion.

Subcontracting (Indicate Type of Contracts)

- Structural steel fabrication - bid, evaluation, award and fabrication.
- Concrete work - bid package, bid, evaluation and award.
- Mechanical work - bid package, bid, evaluation and award.
- Instrument and electrical work - bid package, bid, evaluation and award.
- Other contracts - bid package, bid, evaluation and award.

Construction (All Areas)

- Site preparation.
- Foundations/miscellaneous concrete.
- Structural steel.
- Equipment erection.

- Piping - field fabrication, erection, heat trace and testing.
- Buildings.
- Electrical.
- Instrumentation - installation and check out.
- Insulation and paint.
- Final check out.

Fig 4.3 is a typical summary of the Execution Plan prepared for the case study.
    The observant reader will notice some difference between this and the preliminary execution plan including:

- The construction work hours are higher: 492 versus 466 months.
- Construction duration is somewhat shorter: 8 1/2 versus 9 months.
- Peak staff is higher: 111 versus 103.
- The contingency allowance is lower: 12% versus 20%.

All these changes are typical in the development of any project and result from the preparation of a definitive type estimate after completion of the Phase I Design. In fact, many Project Managers and owners would be extremely happy if the variation between preliminary and definitive estimates were that low.

4.5   Compressing the Schedule

As mentioned in Section 4.4, the minimum capital cost execution plan is rarely consistent with the optimum overall project cost when marketing and other business considerations are taken into account. When the project driving force is the push for early completion, the execution plan must be adjusted through the introduction and implementation of viable schedule reduction measures. The most cost-effective measures are those that can be implemented during the early execution stages when the majority of the activities fall on the critical path.
    The following schedule reduction schemes are some of the most frequently used and are listed in order of the project execution progress, which seems very close to the order of increasing costs and potential drawbacks:

- Approve funding to start engineering and procurement of long delivery equipment prior to formal project approval.
    Potential Drawback - wasted engineering and losing the cancellation charge on early purchases.

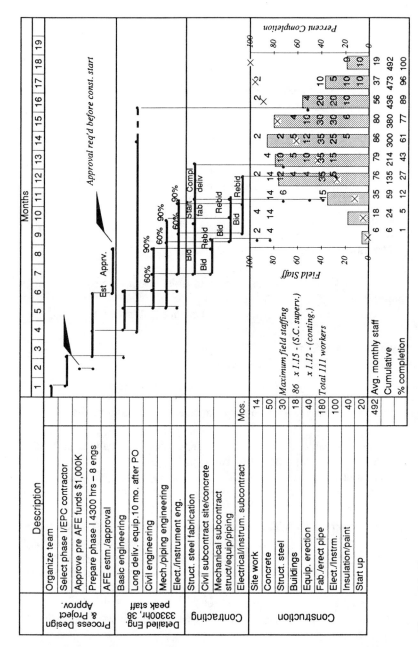

Figure 4.3 Firm execution plan summary.

- When engineering work is being held waiting for decision on alternate schemes, proceed with both in parallel until a decision has been reached.
  Potential Drawback - wasted engineering.

- Work regular overtime in the engineering office.
  Potential Drawback - 3-5% of engineering cost on overtime premium and drop in productivity.

- Place bulk orders of heavier-than-expected structural steel before the design is complete
  Potential Drawback - premium for heavier materials or, worse, finding that the "over design" was not enough.

- Purchase bulk materials from existing stock rather than wait for longer deliveries from the steel mills and/or fabricators
  Potential Drawback - pay higher prices.

- Authorize tanks and vessel fabricators to order materials before drawings approval.
  Potential Drawback - risk of ordering inadequate materials.

- Pay premiums for early delivery of equipment.
  Potential Drawback - higher cost.

- Waive competitive bidding.
  Potential Drawback - lose the advantage of lower competitive prices.

- Break down construction into several discrete subcontracts and plan detailed engineering to support a staged subcontracting plan.
  Potential Drawback - multiplication of subcontracting efforts.
  Lose advantage of better price for larger subcontract packages.

- Start the construction bidding process when engineering is 60% complete. Continue engineering during the bid preparation and ask the two low bidders to rebid on the upgraded design. Continue upgrading the engineering during the rebidding and finally ask the chosen bidder to adjust its proposal to reflect the status of the design at the time of award.
  Potential Drawback - higher subcontracting costs. Possibility of making final award on less than complete engineering and higher than normal field extras.

- Place early bulk orders of material usually provided by subcontractors.
  Potential Drawback - cost of placing additional purchase orders and double handling in the field.

- Execute construction on direct hire basis and/or reimbursable subcontracts.
  Potential Drawback - lose competitive lump sum advantage. Field costs could increase by 15-20%.

- Overstaff the field.
  Potential Drawback - drop in productivity since only a limited number of workers can work efficiently in a given area.

- Execute construction on multiple shifts.
  Potential Drawback - premium pay for night work, drop in productivity and additional fixed costs for preparing the site for night work.

- Work extended overtime.
  Potential Drawback - the combination of overtime premium pay and a substantial drop in productivity could increase field costs by 50 to 60%.

NOTE: The effect of extended overtime on productivity is discussed in some detail in Section 13.9

4.6   Project Coordination Procedure

Although the coordination procedure is not part of the execution plan per se, it is very relevant to it and should be issued, as far as possible, concurrently with it. The coordination procedure shows the project organization chart, defines group and individual responsibilities and sets the basis for financial control. It must include the following information:

- General information:
    1. Project description/location.
    2. Sponsor.
    3. Contractors (if any).

- Group responsibilities:
    1. Division.
    2. Plant.
    3. CED.

      4. Environmental.
      5. Health.

-   Individual responsibilities:
    1. Venture Manager.
    2. Project Manager.
    3. Technical Manager.
    4. Project Engineer.
    5. Plant Representative(s).
    6. Process Engineer(s).
    7. Accountant(s).

    NOTE: All must be identified by name and position within the owner's organization

-   Limits of approval for project expenditures.
-   Scope change procedure and limits of approval.

Appendix C includes a typical organization procedure for an actual project in a chemical company.

CHAPTER 5
ESTIMATING

5.1 Thoughts on Estimating

- In an estimate, the total cost is always more dependable than the cost of each individual portion.

- As mentioned in the previous chapter, estimating and project execution planning go hand in hand; neither can be completed without input from the other. It is the Project Manager's responsibility to insure, through direct and/or supervisory action, that the cross feeding takes place and that both the estimate and the execution plan are thorough and realistic.

- An estimate is not complete without project management/engineering input. The estimator must review the estimate with a project manager/engineer before issuing it. An estimate, to be useful for execution planning, must deal not only with dollars but also with construction and engineering hours. That information is needed to realistically determine duration and staff requirements. An estimating system providing only dollar information may be very useful to management and accountants but has a very limited value to those in charge of planning and executing the project.

- Normally, project managers don't have to prepare cost estimates; estimators will do it for them. However, on small projects, the Project Manager quite frequently must also wear the estimator's hat. Furthermore, on all projects, large and small, the Project Manager must review and approve all cost estimates, thus accepting the ultimate responsibility. In order to discharge that responsibility in an effective manner, the Project Manager must have a good working knowledge of estimating and estimate checking techniques.

    The Project Manager wearing the estimator's hat is not expected to prepare definitive or engineering estimates nor to use computerized techniques but is expected to:

- Have a good understanding of and a feel for costs.
- Develop sufficient estimating skills to be able to prepare conceptual and/or preliminary estimates when so required.
- Be capable of reviewing all types of estimates to ascertain their reasonableness, completeness and accuracy.

- Estimates can be conceptual, preliminary, definitive or engineering, depending on the quantity and quality of information available and the effort applied to their preparation. Estimating techniques range from simple factoring to sophisticated computerized programs, which, if used properly, can produce very accurate estimates. Estimate checking techniques, on the other hand, are probably all manual and the result of experience.
- Accurate estimating is a very important factor to a successful project. A very high estimate could abort a potentially good project while a low estimate, on the other hand, can lead to embarrassing overruns, a lot of aggravation or worse.

   The essential requirements of an accurate estimate are a realistic and thorough scope of work and a sound estimating approach.
- A realistic scope must be based on:

   - Accurate and complete technical information:
     1. Process description and flowsheets.
     2. Drawings.
     3. Specifications.
     4. Site information.
   - Reasonably accurate take-offs:
     1. Materials.
     2. Hours.
     3. Labor Rates.
     4. Indirects.
   - Realistic unit prices/work hours.
   - Thorough and systematic review of all items with potential cost impact.
   - Realistic execution plan/schedule.
- The main ingredients of a sound estimating approach are, in order of importance:

- Common sense and experience.
- Hard work.
- Estimating techniques/systems.

The common sense and hard work applied to thorough project scoping and good execution planning will determine the quality of the estimate regardless of the estimating technique used. They can only be provided by the estimators and/or project managers/engineers. Estimating techniques are, of course, important, especially when they also provide a database. However, they are no substitute for hard work and experience. As mentioned before, estimating techniques range from very simple to very sophisticated. There is a niche for each technique. However, simple ones are normally quite adequate for small projects.

- Sometimes it seems that accuracy has little to do with the effort and sophistication of the resources applied to the preparation of estimates. Occasionally a 15 minute, back-of-an-envelope estimate turns out to be more accurate than a detailed estimate prepared after weeks of engineering using complicated and expensive techniques. However, these are extreme cases, and a back-of-an-envelope estimate should not be relied upon for appropriation nor for cost tracking and/or control.
- The estimate must provide sufficient information to establish its credibility and contain sufficient details to permit cost and progress tracking. To accomplish this with reasonable accuracy and minimum effort and paperwork should be the prime goal of estimators and project managers/engineers. While a minimum of detail is required even in conceptual and preliminary estimates, extensive details are overwhelming without adding any accuracy to the estimate.

  Factored estimates, especially when used for repeat plants, can be accurate and adequate for appropriation purposes but don't provide the information required for cost and progress tracking. They should be followed as soon as possible with more detailed estimates containing sufficient and suitable information for tracking. On the other hand, the detailed estimates normally prepared by engineering contractors after a certain amount of design work are time consuming and expensive and should be used sparsely.
- Some owners prefer to base the formal approval of a project on a detailed cost estimate prepared by an engineering firm and are willing to release funds to do the front-end design required for the estimate. Others, even if they would like to have the cost and progress tracking capability, are very hesitant to start spending on engineering before formal project approval.

The Project Managers then find themselves between a rock and a hard place. Although there is nothing they can do to change management policies they are still expected to come up with good quality appropriation estimates no matter how meager the available resources are.

THE SEMI-DETAILED ESTIMATING TECHNIQUE PROPOSED IN CHAPTER 13 PROVIDES THE OWNERS' PROJECT MANAGERS WITH A SIMPLE WAY TO PREPARE QUALITY ESTIMATES WITHOUT RESORTING TO CONTRACTORS OR TO COMPLI-CATED TECHNIQUES.

5.2   Estimate Definitions

Cost estimates are categorized according to the quality of the basic process information available rather than on the meticulousness of the estimating procedure and the details included in the estimate.

A PRECISE ESTIMATE IS NOT ALWAYS AN ACCURATE ESTIMATE.

For example, under certain circumstances, a series of theoretical and/or test tube chemical reactions could be the basis of a conceptual but impressive-looking design package with enough information to prepare a detailed estimate. However, regardless of the details, the resulting estimate cannot be anything but conceptual.

Order of Magnitude Estimate

These estimates provide only ball park costs and are sometimes required in the very early stages of project development as decision-making tools. The basis could be no more than a given capacity of a generic product, the estimating procedure, a simple search of existing literature and, if required, the application of the proportioned estimating method.

Conceptual Estimates

Conceptual estimates are typically prepared during the advanced R&D and early process design stages. They are normally used to build the preliminary project economics, prepare preliminary execution plans and develop cost estimates for the preparation of a complete process design (Phase I) and a definitive cost estimate. The Phase 0 design package, described in Chapter 3, contains all the information required for conceptual estimates.

Preliminary Estimates

These estimates are prepared when the basic process design is essentially complete and the scope of the off-site and waste treatment facilities has been established but not completely defined. Preliminary estimates are normally used to prepare firm project execution plans, to complete the project economics and, frequently, for appropriation purposes. The Phase I design package, described in Chapter 3, contains the necessary information for the preparation of a preliminary estimate. In fact, a thorough Phase 0 package could also be used for this purpose.

Definitive Estimates

The definitive estimates are prepared after the project has been totally defined, including off-sites, waste treatment, site conditions, equipment arrangements, etc. This stage of definition requires the involvement of an engineering contractor to complete the so-called basic engineering referred to in Chapter 7.

It is advisable to complete the definitive estimate before the fabrication of expensive equipment and start of construction as a protection against major, unexpected cost variations which could have a negative influence on project approval.

Engineering Estimates

They are prepared at various points during the detailed engineering mainly as a control and cost tracking tool.

5.3 Estimating Methods

The estimating methods normally employed fall in five categories.
- Proportioned.
- Factored.
- Computerized simulations.
- Detailed.
- Semi-detailed.

Proportioned Method

The proportioned method is used to estimate the cost for a new size or capacity from the actual cost of a similar completed facility of known size or capacity. The relationship has a simple exponential form:

$$\text{Cost } 2 = (C_2/C_1)^n \times \text{Cost } 1$$

Where: C_1 = known capacity or size

C_2 = new capacity or size

n = cost capacity factor

This method should be used only for order of magnitude estimates.

Since this formula is assumed for the same time period and location, the results must be adjusted by using an escalation index such as Chemical Engineering Plant Cost Index.

The cost capacity factor (n) has an average value of 0.6 for most plants and equipment but can vary over a wide range. The factor can be obtained from published data or historical records.

Factored Method

In the factored method, the total cost, broken down by the main accounts, is arrived at by multiplying the cost of equipment by certain empirical factors which take into account whether the process involves handling fluids, solids or a mixture of both. Since the original factors were developed by H.J. Lang, the method is known as the Lang Factor method. However, through the years, contractors and owners have developed their own factors to reflect their particular experiences.

Tables 5.1, 5.2 and 5.3 illustrate a set of factors used by one chemical firm for fluids-handling and solids-handling plants. The factors and ratios used to break down the total cost into the various direct and indirect accounts are based on a 100% subcontract construction where most of the labor-related field indirects are collected with the subcontract labor costs and the "field indirects" account refers to only the construction management and the site general conditions costs. Field indirects are discussed in more detail in Section 13.9.

The ranges of the factors recommended for the various cost accounts are very wide and the accuracy of the estimate of each one will be much lower than the accuracy of the total cost. Additionally, the resulting estimate would not be suitable for cost tracking.

Some of the driving forces determining the location of each factor within the ranges are:

- **Process** - Dry process usually has more equipment, less piping, less instruments, less engineering, more electrical and about the same civil and structural requirements.

- **Toxicity** - Highly toxic or hazardous processes can drive all categories higher, but usually piping, instrumentation and equipment are the most affected.

Table 5.1 Typical Factors - All Fluid Plants

	Range % Equipment	Usual factor Fraction of Equipment		
		Material	Labor	Total
Direct Costs				
Equipment	–	**1.00**	**0.20**	**1.20**
Site development	4-8	0.01	0.05	0.06
Concrete	10-50	0.05	0.25	0.30
Structural steel	20-45	0.20	0.15	0.35
Buildings	1-15	0.06	0.04	0.10
Piping	120-200	0.50	1.10	1.60
Electrical	20-50	0.12	0.20	0.32
Instrumentation	25-75	0.30	0.15	0.45
Insulation	5-30	0.06	0.04	0.10
Painting	2-8	0.01	0.03	0.04
Fire protection	5-15	0.05	0.05	0.10
Subtotal		**1.36**	**2.06**	**3.42**
Total Direct Costs		**2.36**	**2.26**	**4.62**
Indirect Costs				
Home office		20% of directs		0.93
Field		20% of labor		0.45
Total Indirects				**1.38**
Total Installed Cost (TIC)				**6.00**

Home Office   0.93 / 6.00 = 0.16 TIC
Labor   2.26 / 6.00 = 0.38 TIC

Table 5.2 Typical Factors - Fluid / Solid Plants

	Range % Equipment	Usual factor Fraction of Equipment		
		Material	Labor	Total
Direct Costs				
Equipment	–	**1.00**	**0.20**	**1.20**
Site preparation		0.01	0.05	0.06
Concrete	12-55	0.06	0.29	0.35
Structural steel	25-50	0.23	0.17	0.40
Buildings	1-15	0.06	0.04	0.10
Piping	60-160	0.35	0.65	1.00
Electrical	25-60	0.15	0.25	0.40
Instrumentation	20-70	0.27	0.13	0.40
Insulation	5-15	0.05	0.03	0.08
Painting	3-8	0.02	0.03	0.05
Fire protection	5-15	0.05	0.05	0.10
Subtotal		**1.25**	**1.69**	**2.94**
Total Direct Costs		**2.25**	**1.89**	**4.14**
Indirect Costs				
Home office		20% of directs		0.83
Field		20% of labor		0.38
Total Indirects				**1.21**
Total Installed Cost (TIC)				**5.35**

Home Office   0.83 / 5.35 = 0.16 TIC
Labor   1.89 / 5.35 = 0.35 TIC

Estimating   63

Table 5.3 Typical Factors - All Solids Plants

| | Range % Equipment | Usual factor Fraction of Equipment | | |
		Material	Labor	Total
Direct Costs				
Equipment	–	1.00	0.20	1.20
Site preparation	4-8	0.01	0.05	0.06
Concrete	15-60	0.06	0.32	0.38
Structural steel	25-60	0.25	0.20	0.45
Buildings	1-15	0.06	0.04	0.10
Piping	30-80	0.20	0.40	0.60
Electrical	25-60	0.15	0.25	0.40
Instrumentation	20-70	0.27	0.13	0.40
Insulation	5-10	0.04	0.03	0.07
Painting	4-8	0.02	0.03	0.05
Fire protection	5-15	0.05	0.05	0.10
Subtotal		1.11	1.50	2.61
Total Direct Costs		2.11	2.26	3.81
Indirect Costs				
Home office		20% of directs		0.76
Field		20% of labor		0.34
Total Indirects				1.10
Total Installed Cost (TIC)				4.91

Home Office   0.76 / 4.91 = 0.15 TIC
Labor   1.70 / 4.91 = 0.35 TIC

- **Corrosiveness** - Highly corrosive processes drive piping, equipment and instrumentation costs up.
- **Capacity** - May dictate oversized equipment, etc., for increased production. This would tend to drive civil and structural accounts higher.
- **Site Conditions** - If unfavorable, would drive the civil and structural accounts higher.
- **First-of-a-Kind Unit** - Would tend to drive engineering disproportionately higher.
- **Weather/Location** - Drive labor overheads higher.

In general, the factored method is recommended only for conceptual estimates or those cases when reliable information is available on the actual cost of similar projects.

Computerized Simulations

This method is a valuable tool for the estimator. The system calculates the purchased price for each piece of equipment as well as simulates (based on preprogrammed models) quantity take-offs to generate field material and field labor for installation of each equipment item as well as the entire project. Engineering, overheads and fees are also calculated. Certain minimum information must be input to the computer for equipment costs. More data given per piece of equipment results in a more accurate cost estimate of the equipment and project as a whole.

The computerized method is an excellent tool for conceptual and even preliminary estimates but should not be used for definitive estimates unless applied by experienced estimators fully familiar with the algorithms and unit prices built into the program for developing takeoffs and calculating material costs and labor hours.

Detailed Method

The detailed method of estimating is usually employed by engineering firms and/or contractors. This method is, as its title implies, much more time consuming than the previous methods.

Take-offs are made from actual drawings. Quotes are obtained on major equipment and subcontracts. Detailed bulk material take-offs are made and priced. All the field labor hours are estimated based on quantity take-offs. The engineering hours and costs, supervision and field expenses and other items, such as freight, duty and taxes, insurance, etc., are all detailed and priced.

The preparation of a detailed estimate involves a team effort. There must be an extensive review and checking process before the estimate can be completed to ensure that every item is accounted for.

This method is best suited for contractors preparing estimates to bid work on a lump sum basis. The additional time and cost required could be difficult to justify even for the preparation of appropriation estimates.

Semi-detailed Method

The semi-detailed method described in Chapter 13 can be as accurate as the detailed method at a fraction of the time and cost. It is essentially based on equipment count and preliminary take-offs of comprehensive, yet comprehensible, units priced with equally comprehensive unit prices and unit hours.

While a detailed estimate could not be prepared without the benefit of some detailed engineering, a good semi-detailed estimate can be prepared with the information available at the completion of the Phase I design.

- Equipment list and specifications.
- Complete P&ID's.
- Plot plans.
- Preliminary equipment arrangements.

The quality of a semi-detailed estimate will, of course, be enhanced if it is prepared at the completion of the basic engineering.

5.4   Anatomy of an Estimate

General

The cost estimate must be consistent with the Project Execution Plan and reflect the execution approach.   If the approach is to use multiple construction contractors, the estimate backup must contain sufficient details so that it can be recast with minimum effort to reflect the scope and cost of each contract.

An appropriation estimate, whether prepared in-house or by outside contractors, must clearly identify the project, its location and the quality of the estimate.   It should be issued in the form of a formal, stand-alone, report or package addressing and documenting the following areas:

- Cost summary (direct and indirect).
- Basis of estimate/scope definition.

- Execution approach.
- Milestone schedule.
- Accuracy and risks.
- Estimating technique/approach.
- Backup.

Although the extent of documentation required will be different for each project, it should be generally consistent with the succeeding paragraphs.

- The cost summary must contain sufficient information to allow management to calculate and analyze the traditional factors and ratios to ascertain their consistency with industry standards and past experience. The following information must be included:

  - Breakdown between direct and indirect costs.
  - Breakdown of directs by equipment and commodities, materials, labor and subcontracts.
  - Equipment count by type.
  - Major take-offs - total cubic yards of concrete, tons of steel, number of motors, area of buildings, etc.
  - Estimated engineering hours.
  - Estimated field hours including subcontracts.
  - Breakdown of indirects by engineering, field indirects, other costs, escalation and contingency.
  - Expense items.

  Table 5.4 is a typical example of a cost estimate summary. The suggested allocation of costs to the various accounts is discussed in Section 5.5.

- Every cost estimate is based on a specific scope which may range from a single phrase (i.e., "One 20 MPY Oxinitro Plant in Brazil") to an engineering package complete with Approved for Construction drawings and specifications. For the size of projects discussed in this book, the scope would usually be based on in-house Phase 0/I design. When the available information is incomplete, the Project Manager must make conservative educated assumptions to prepare the estimate. The basis of the estimate, as well as the assumptions made, including any specific inclusions and/or exclusions must be clearly defined and documented as part of the estimate package.

Table 5.4 Cost Estimate Summary Case Study

Description	Qty	Unit	Work Hours	Mat'l by Owner	Subcontract Labor only	Subcontract Mat'l & Labor	Total
					Thousand $		
Direct Costs							
Equipment	62	Ea	7,200	1,431	286	189	1,906
Freight 3%				80			80
Piping	14,000	LF	29,200	700	1,210	145	2,055
Instrumentation	–	Lot	8,300	752	–	460	1,212
Electrical	34	Lot	8,000	372	336	–	708
Site preparation	2,400	SY	900	0	0	90	90
Fire protection	–	Lot	1,500	0	0	145	145
Concrete	700	CY	8,600	0	0	440	440
Structural steel	190	Tn	4,400	421	198	0	619
Buildings	280	SF	3,000	–	–	300	300
Insulation	–	Lot	5,100	0	0	380	380
Painting	–	Lot	1,300	0	0	65	65
Demolition/ Relocation	–	Lot	0	0	0	0	0
Total Directs			**77,500**	**3,756**	**2,030**	**2,214**	**8,000**

Indirect Costs

Field Indirects 20% of 2030 + 12% of 2214		670
Engineering contractor H.O. cost - 30,000 hrs @ $60/hr		1,800
CED costs - including phase I		450
Taxes		0
Spare parts	10% of equipment cost	140
Start up	7% of labor cost (5% labor; 2% mat'l)	140
Total Indirects		**3,200**

Total Directs + Indirects	**11,200**
Escalation 3 %	330
Target Estimate	**11,530**
Contingency = 20%	2,270
Total Installed Cost	**13,800**

- The execution approach and contracting strategy will affect engineering costs, labor productivity and field indirects. Since the cost impact could be substantial, the estimate package must include at least a simple description of the execution plan.

  The time frame in which a project is expected to be executed impacts several important cost areas:

  - Labor rates.
  - Contract negotiations.
  - Labor availability.
  - Contractor's work load.
  - Escalation.
  - Plant tie-ins and shutdown work.
  - Weather effects.

  Although fast tracking could ultimately have a beneficial impact on the project, its cost will always be higher than optimum. If the project is to be fast track, it must be taken into consideration when the cost estimate is prepared.

  All estimates must be accompanied by at least a simple milestone schedule to highlight the important dates. When variations in the project time frame could have a substantial impact on cost, the estimate package must also include a discussion on the subject.

- Determining the accuracy of an estimate and the risk of variations is an extremely difficult proposition. Trying to quantify them is like playing Russian roulette. However, no matter how arduous the task, the questions must be addressed as part of the estimate package. Several methods are available to mechanize the determination of estimating accuracy. However, ultimately, it is up to the Cost Engineer and/or Project Manager to apply their experience and common sense and evaluate the accuracy of the estimate through the analyses of the entire estimating process:

  - Details of scope.
  - Engineering input.
  - Source of equipment and materials prices.
  - Estimating techniques utilized.
  - Take-off allowances.

- Source of labor rate and productivity data.
- Viability of construction approach.

The degree of accuracy and the level of risk can be controlled through the judicious addition of resolution allowances to the various accounts and escalation and contingency to the estimate total. The application of these cost adjustment factors is discussed in Section 13.10.

All information used and/or developed for the estimate must be documented and kept at least until the project is completed and audited. This includes drawings, take-offs, preliminary sketches, vendor quotations, etc. The backup information is required for cost and scope tracking and control as well as final cost analysis.

Breakdown

With the exception of factored estimates, all others, from conceptual to engineering, must have the potential of being broken down into discrete components that can be related to the project schedule. The estimate breakdown is a very important tool for project planning as well as scope, cost and progress tracking. In the case of conceptual and preliminary estimates, the breakdown is required to prepare the master schedule and project execution plan (see Chapter 3). In the case of definitive and engineering estimates, it is essential to the preparation of the cost tracking and construction progress monitoring systems. The extent of the breakdown is dependent on the availability of time and scope details. However, a thorough one must be provided with appropriation estimates or when there is a high probability of executing the project.

> THE COMPONENTS OF THE ESTIMATE BREAKDOWN MUST BE QUANTIFIABLE AND SMALL ENOUGH TO BE EASILY COMPREHENDED AND EVALUATED BY SIMPLE OBSER- VATION WITHOUT GOING TO THE DETAILS OF COUNTING AND TRACKING "NUTS AND BOLTS".

The estimate can be broken down to various levels of detail consistent with the:

- Availability of information.
- Magnitude of the project.
- Need to keep components small enough for evaluation by simple observation.

The first level of breakdown is obviously the one shown in the Cost Estimate Summary. This level is sufficient for the preparation of preliminary execution plans and master schedules.

The second level would be the breakdown of all first level components by process areas or steps, utility areas and common facilities. This breakdown is required for more advanced planning.

The actual cost and progress tracking, especially on larger projects, require further breakdown of cost items, such as:

- **Equipment Cost, Erection, Insulation** - by items.
- **Site Preparation** - by type of work (fences, roads, railroads, site clearance, piling, etc.)
- **Concrete and Steel** - by physical area and/or equipment item.
- **Piping/Insulation** - by system and/or line.
- **Electrical** - by motors, feeders, substations, illuminated areas, etc.
- **Instrumentation** - by systems, equipment item and/or "balloon count".

More details on cost breakdowns are included in Chapter 10.

Some engineering and construction firms, as well as some of the computer programs in the market, carry the breakdown to the level of nuts and bolts (square feet of forms, pounds of rebar, field welds and bolt-ups, individual instruments, etc). That level of detail might be important to some nitpicking accountant or perhaps to an efficiency expert in the field, but has no use for the Owner's Project Manager interested in the overall picture.

Breaking down an estimate requires a certain degree of engineering and materials take-offs. In the case of definitive and engineering-type estimates, this will usually be done by an engineering contractor and the details will be more than adequate for cost tracking. In the case of conceptual and preliminary estimates prepared in-house, this work must be done by the Owner's engineers. In the specific case of small projects, the Project Manager is expected to perform the required conceptual engineering and preliminary take-offs as part of the estimating work. The proposed semi-detailed estimating method described in Chapter 12 provides tools to prepare quality estimates with sufficient breakdowns for execution planning, cost tracking and progress monitoring.

5.5   Cost Allocation

Although most owners and contractors break project costs along similar lines, there are always differences that could distort the analyses of cost factors and ratios. To avoid distortions and the resulting erroneous analyses, each Owner should standardize the manner in which costs are allocated in the estimate and insist that contractors recast at least their cost summaries following the same system. This section is intended as a guideline for project costs allocation to be used by the project managers as well as outside contractors estimating projects for them when there is no standard. The guideline is consistent with the subcontracted construction approach recommended in Chapter 6 and the estimating method described in Chapter 13.

The very basic cost breakdown is by directs and indirects. The directs include all materials that physically go into the project and the cost of the labor required to install it. The indirects include engineering, field indirects, spare parts, taxes, start-up, etc.

The suggested basic allocation of directs, applicable to all the estimate accounts is by:

- Materials.
- Subcontracted labor.
- Material and labor subcontracts.

The materials column includes all the equipment and materials usually purchased by the Owner and/or general contractor - to be installed by the subcontractors. The subcontracted labor column includes all costs, direct and indirects, incurred by the subcontractors for the installation of these materials. Construction equipment rentals as well as contractor overhead and profit must also be included. The accounts traditionally treated in this manner are:

- Equipment.
- Piping.
- Instrumentation.
- Electrical.
- Structural steel.

The subcontract column includes those items where a vendor or a subcontractor supplies both materials and erection. The cost includes materials, labor, labor indirects, construction rental equipment, and contractor/vendor overheads and profits. The following accounts are traditionally executed and estimated on a subcontract basis.

- Vendor erected equipment.
- Site preparation.
- Sewers.

- Fire protection system/fireproofing.
- Concrete.
- Buildings.
- Painting.
- Insulation.
- Dismantling.

The estimated construction hours, even those related to subcontract work, must also be shown in the estimate cost summary. That information is vital to prepare the execution plan and the construction monitoring systems.

The direct costs must also be allocated by accounts, and each account, of course, broken down by materials, labor and subcontracts. The succeeding paragraphs discuss the allocation to each of the accounts.

- The **Equipment** account could be broken down by types of equipment and, in definitive and engineering estimates, even by equipment item. Electrical equipment, transformers, switch gear, etc., should be included in the electrical account.

  Freight must be included with the equipment cost either as a separate line item or included with purchase cost of each item.

- The **Piping** account covers all process and utility piping, as well as chutes and ducts, above and underground, including:

- All piping materials and welding supplies.
- Hangers and supports.
- Receiving/storing/handling.
- Shop and field fabrication.
- Erection.
- Cleaning and testing.
- Tie-ins.
- Safety showers and eye washes.
- Heat tracing - both steam and electric.

The following items should be excluded from the piping account:

- Sewer (with Sewers Site Preparation).
- Storm drains (with Site Preparation).
- Fire protection systems (with Fire Protection).

- Engineered pipe supports (with Structural Steel).
- Instrument air from distribution point to individual instrument (with Instrument).
- Utility supply to battery limits (with Site Preparation).
- Pipe racks and sleepers (with Structural Steel or Concrete).

When the piping installation is subcontracted, the subcontractor frequently supplies the bulk materials. Even if that is the case, the account should show the breakdown between material and labor to permit the analysis of the material/labor ratio.

- The **Instrumentation** account includes:

  - All field and panel instruments and installation materials.
  - Control panel/TDC/PLC.
  - Receiving/storing/handling.
  - Calibration.
  - Installation.
  - Loop checkout.
  - Instrument air piping from distribution point to the individual instrument.
  - Electrical wiring from the individual instruments to control center (TDC).

Normally, the installation of instruments is subcontracted with the subcontractor supplying the installation materials. Those materials should be included together with the subcontract labor in the subcontract column.

- The **Electrical** account should include:
  - All electrical equipment:
    1. Unit transformers and associated switch gear.
    2. Lighting transformers.
    3. Motor control centers/breakers.
    4. Local panels.
  - All materials required for high and low voltage distribution from the main substation (excluded) down to the last lighting fixture.
    1. Poles, towers and miscellaneous supplies.
    2. Conduit and trays.

3. Wiring and bus ducts.
4. Junction boxes.
5. Terminals and connectors, push buttons, etc.
6. Lighting fixtures and receptacles.
7. Lighting panels.
8. Welding and utility receptacles.
9. Grounding.
- All labor required to install the above materials.

The following items should be excluded from the electrical account:

- Electric heat tracing (with Piping).

- Main substation and related switch gear (see note below).

- Instrument wiring (with Instrumentation).

- Lighting of buildings contracted in design-build basis (with Buildings).

- Foundations and structural steel supports for the electrical equipment (with Concrete and Structural Steel).

When the electrical installation is subcontracted, the subcontractor frequently supplies the bulk materials. If that's the case, the materials should still be shown in the materials column to permit the analysis of the Material/Labor ratio.

NOTE: Usually only true grass-roots projects require a high voltage (main) substation. Most of the projects contemplated in this guideline would be built on existing developed sites and will require only unit substations (480v). Including the main substation, if required, in the electrical account will distort the traditional factors and ratios used in checking the estimate. The main substations and high voltage distribution should be included with the site preparation account or as a stand-alone item.

- The **Site Preparation** accounts should include:

    - General excavation, backfill, shrubbing and grading.

    - Piling/soil stabilization.

    - Roads, fences, railroads, parking areas, culverts, etc.

    - Retention ponds and lagoons.

    - Spillways/rip-rap.

    - Storm and sanitary sewers.

-   Underground process sewers to off-plant treatment facilities.
-   Utility supplies:
    1. Power supply feeders.
    2. Water/gas/other.

The following items should be excluded:

-   Process sewers to in-plant treating facilities (with Piping).
-   Pavement around process areas (with Concrete).
-   Excavation and backfill related to equipment and structural foundations (with Concrete).

-   The **Fire Protection** account should include:
    -   Fire water distribution systems including hydrants, monitors, etc.
    -   Sprinkler/foam systems.
    -   Valve and hose houses.
    -   CO_2/halon system.
    -   Fireproofing.

The following should be excluded:

-   Fire water storage tanks and pumps (with Equipment)
-   Fire water lagoons (with Site Preparation)
-   Fire protection of buildings contracted on design-build basis (with Buildings)

-   The **Concrete** account should include:
    -   Concrete tanks/basins/sumps.
    -   Pile caps.
    -   Ground and elevated slabs.
    -   Paved process areas.
    -   Dikes, fire walls and trenches.
    -   Excavation and backfill associated with the above.

The following should be excluded:

- Foundation for buildings contracted on a design-build basis (with Buildings).
- Main substation foundations (with Site Preparation).

- The **Structural Steel** account should include:

  - All equipment supporting structures, including platforms, stairs, ladders, railing, etc.
  - Pipe racks.
  - Miscellaneous structures attached to equipment unless provided as part of the vendor's supply.
  - Truck and railcar loading and unloading platforms and arms.

  The following items should be excluded:

  - Structures associated with buildings contracted on a design-build basis (with Buildings).
  - Standard hangers and supports for piping, instrumentation and electrical (with corresponding accounts).

  Normally the structural steel is fabricated under a purchase order and erected on a subcontract basis.

- The **Buildings** account should include:

  - Shelters - roof and walls added to process structures.
  - Control and motor control rooms - usually in the same building.
  - Warehouses, shops, change rooms, offices, etc.
  - All HVAC and lighting associated with the above.

- The **Painting and Insulation** accounts should include:

  - Equipment painting and insulation.
  - Piping painting and insulation.
  - Structural steel painting.

- The **Dismantling** account should include:

  - All equipment, piping and other removals and/or relocations required to prepare an existing area for the installation of the facility.

- The **Field Costs** cover all costs associated with construction management and are discussed in detail in Sections 13.9. They do not cover the indirects associated with subcontracted labor included in the labor column.

- **Contractor Home Office** covers all engineering costs including contractor's overhead and fees. When engineering is done in-house, the costs also may be included in this account and identified as a separate item from contract engineering.

- The **Start-up** account is intended to cover items like:

  - Minor changes required to start up the facility.
  - Correcting minor errors.
  - Supply standby personnel during start-up.

- All other indirect accounts are self-explanatory:

  - CED costs.
  - Taxes.
  - Spare parts.
  - Royalties.
  - Catalysts and chemicals.

Finally, all estimates must include, as part of the individual accounts, a resolution allowance, plus escalation and contingency added to the bottom line. These cost items are discussed in the next section.

5.6   Adjustments

Resolution Allowance

When an estimate is based on firm vendors' quotes and materials take-offs, experience shows that, invariably, some items affecting cost will be overlooked and left out of the estimate. To compensate for these oversights, it is appropriate to include in the estimate allowances for the unlisted items. These are called Resolution Allowances.

The resolution allowances are intended to cover costs that experience shows will be incurred during the course of the project and, therefore, should not be considered contingency but should be included with the various cost accounts. The amount of resolution allowance applied to each account is dependent on the stage of development of the information used to determine equipment costs and prepare materials take-offs. However, in the final reckoning, they reflect the judgment and experience of the Cost Engineer and/or the Project Manager. Chapter 12 includes a guideline recommending the amount of resolution allowance that should be applied to the various cost accounts at different stages of engineering.

Escalation

Estimates are prepared based on prevailing prices at the time of the estimate while actual expenditures will be incurred some time in the future at prices pushed up by inflation. The cost estimate must, therefore, be adjusted by an escalation factor to account for inflation and bring it as close as possible to the actual cost at the time of completion. Although the techniques employed to arrive at the escalation factor will vary, all are based on published inflation indexes and projections.

The simplest method, quite adequate for small projects, is to apply the current inflation projection to the mid-life of the project; i.e., if the estimated duration is 14 months and the projected annual inflation rate is 6%, the escalation adjustment will be:

$$(14 \text{ months} \div 12 \text{ months/year}) \times 0.5 \times 6\% = 3.5\%$$

On large, extended projects where the escalation allowance could become a substantial number, it is more appropriate to calculate the escalation for each of the major cost accounts based on the projected inflation rate and the project schedule. By doing so, every cost account will be escalated to the time in which most of the actual expenditure is projected.

Special care must be taken in escalating labor costs since the rates are determined by labor agreements lasting two or three years, and it is quite possible that no change in rate occurs during the project.

Contingency

No matter the high degree of scope definition or the sophistication of the estimating techniques, a cost estimate is still a forecast of events to come and as such it cannot be perfectly accurate. The contingency is a bottom line addition to the cost estimate to convert it to the most likely final cost.

It is intended to cover:

- Take-offs and arithmetical errors and oversights.
- Variances in equipment and material prices.
- Variance in labor rates and productivity.
- Schedule delays.
- Changes in economic and environmental conditions.
- Changes in execution approach.
- Miscellaneous design changes not affecting the basic scope, including:
  1. Addition of equipment and instrumentation.
  2. Size changes in equipment, piping, wiring.
  3. Changes in pipe and electrical tray routing.
  4. Moderate changes in environmental regulations.
  5. Lower than expected labor productivity.

The contingency is not intended to cover changes affecting the basic project parameters as defined and accepted in the scope, such as:

- Capacity.
- Raw materials and products mix, quality and specifications.
- Facility life expectancy.
- Major changes in environmental regulations.
- Force majeure events (strikes, acts of God, etc).

As in the case of the resolution allowances, the magnitude of the contingency is dependent on the stage of development of the information used in the estimate and reflects the judgment and experience of the Cost Engineer and the Project Manager. Contingency is a forecasting and control tool that is never to be treated as a "checking account". It should be re-evaluated every time the project is reestimated and adjusted according to the prevailing circumstances.

5.7 Checking Criteria and Guidelines

Regardless of who prepared the estimate, the Project Manager is ultimately responsible for the project cost and must take an active role in the monitoring, as well as in the review and approval of all estimates. This applies to in-house as well as contractors' estimates.

The involvement must be far more extensive than the mere review of the final product. The Project Manager must also review the basis of the estimate and the estimating approach to insure that the estimating efforts are commensurate with the information available and that the product will include sufficient details to satisfy the project's further need for planning, cost tracking and progress monitoring.

Finally, the Project Manager must document (or insure that the Cost Engineer does) the estimate review in sufficient detail to be useful for future reference. The review should be part of the estimate package discussed in Section 5.4: Anatomy of an Estimate.

The determining factors in the quality of an estimate are:

- The firmness of the scope and soundness of the design basis.
- The extent of the engineering developed for the estimate.
- The thoroughness of cost items identification (take-offs).
- The validity of unit prices/man-hours used to build up the total cost.

To assess the firmness of the scope and design basis, the Project Manager must question the project sponsors as well as the engineers (process, production, specialists) responsible for the development work to ascertain that they have done their homework and to promote further thought.

The Project Manager must also investigate the extent of the engineering work performed prior to the estimate and the overall quality of the documentation provided to the estimators.

- PFD's and P&ID's:
  1. Preliminary.
  2. Reviewed/approved by Owner.
  3. Approved for design and/or construction.
- Layouts/equipment arrangements:
  1. Preliminary.
  2. Reviewed by owner.
- Detailed engineering by discipline:
  1. Conceptual.
  2. Preliminary.
- Estimating tools.

As discussed in Section 5.4, the breakdown of cost items (take-offs) can go down to the level of nuts and bolts (detailed) or be based on larger, more comprehensive units (semi-detailed). Both approaches are valid provided that the scope of the unit prices corresponds to the units and that adequate resolution allowances are included to compensate for omissions and oversights. The estimate review must address the breakdown philosophy to ascertain that the unit prices and resolution allowance are consistent with the details of the take-offs.

The Project Manager must spend time with the estimators to understand the overall estimating technique, discuss the take-off methodology and be satisfied that all cost items have been considered. The Project Manager should also perform quantity spot checks and challenge any variations greater than 20%.

- **Concrete and Structural Steel** quantities can be easily checked with the quick estimating procedures in Chapter 13.
- **Instrumentation and Electrical** take-offs can be checked by quick balloons and motors count on selected areas.
- **Piping** take-offs can be checked by quick count of lines and valves on randomly selected P&ID's.

After the accuracy and thoroughness of the take-offs have been established, the validity of the materials and labor cost information must be determined. This can be done by determining the source of the data used by the estimator and checking it against an independent source and/or the unit prices/hours in Chapter 12. Any variation of more than 15-20% should be challenged and investigated.

Even if the validity of the basic unit costs and hours has been established, the Project Manager must ascertain that the composition of each unit price is consistent with the scope of units; i.e., if the unit is cubic yard of concrete and includes excavation, rebar and forms; the cost and hours per cubic yard must include a reasonable prorata of all the cost components. Chapter 13 provides enough data to make this analysis.

The unit hours in the estimating procedures, in Chapter 13, are based on so-called Gulf Coast productivity which tends to be conservative since it is mainly based on a direct-hire situation and unionized labor practices. They very seldom find their way into an estimate but are adjusted, up or down, to reflect specific project requirements, contracting strategy and general business conditions as well as local labor availability and atmosphere. It is up to the Project Manager to ascertain the validity of the labor productivity used in developing construction man-hours. Labor productivity is discussed in some detail in Chapter 13.

Indirect costs must be investigated separately to determine whether they have been independently estimated or simply factored from the direct materials and labor costs.

- Estimating **Field Indirects** as a factor of labor costs is quite appropriate providing the factor recognizes the difference between direct labor costs and subcontract labor costs, which already include a large fraction of indirects as well as the impact of fast track schedules on field indirects.
- On the other hand, estimating **Home Office** (engineering) costs as a factor of directs is a dangerous practice. Engineering costs should be estimated separately or related to equipment type and count.

The Project Manager must also make an in-depth review of the resolution and contingency allowances included in the estimate, understand the estimator's rationale and be satisfied that they are consistent with the estimating approach and not over- or understated.

Finally, the Project Manager should make, as a check, rough estimates of the equipment-related accounts (instrumentation, electrical and engineering) using the costs data and guidelines included in Section 13.11.

The following paragraphs list some of the cost items often overlooked in estimates:

- "Spaghetti" piping:
    1. Steam tracing supply and condensate return.
    2. Instrument pneumatic supply.
    3. Safety showers and eyewash stations.
    4. Utility stations.
    5. Inert gas blanketing.
- Freight:
    1. Equipment.
    2. Miscellaneous materials.
- Bulk materials discounts.
- Spot overtime.
- Utility tie-ins.
- Capability of existing utilities to support new project requirements.
- Fast tracking costs:
    1. Regular overtime.
    2. Field productivity losses.
    3. H.O. productivity losses.
    4. Over-ordering of materials.
    5. Double bidding subcontracts.
    6. Vendor's overtime/bonuses.
    7. Additional field extras.
    8. Premature start of field work.

The estimating checklist in Appendix D should be a very useful tool for checking the completeness of estimates.

CHAPTER 6
CONTRACTING

6.1 Overview

Contractor selection is probably the activity that has the most lasting effect on project execution. A poor design or a bad estimate can always be revised and, if caught in time, the effects minimized. Changing a contractor after the work has started is a very difficult proposition that always has a negative impact on project cost and schedule. The most important consideration affecting contracting is whether the project will be executed as a large project or as a small project. This decision must be made by management prior to the start of the contracting activities.

The considerations involved in contracting for large and small projects are different. Contractors involved are also different. Naturally, the factors affecting contractor qualification and selection must also be different. Although this book addresses the execution of small projects, the material and guidelines presented in this chapter are appropriate for all projects regardless of size.

The prime responsibility of contracting rests with the Contract Engineer working for CED. However, the Project Manager must live with the selected contractor and make it perform. The Project Manager must have an active participation in contracting activities from the development of the strategy to the contract award and be directly responsible for all technical aspects of the contracting process - bidder qualification criteria, scope of work, bid evaluation criteria and the technical evaluation of bids.

After contract award, it is the Project Manager's responsibility to implement all contractual provisions and have a complete understanding of the contract in general and, in particular, of those clauses asserting owner's rights of technical approvals and control of the purse strings for reimbursable charges.

6.2   General Considerations

The contract is one of the most important project documents and regulates the relation between Owner and contractor(s). It must clearly define the what, the who, the how, the when and the how much.

WHAT: The scope of the work must be clearly and accurately described.

WHO: The responsibility of both owner and contractor must be clearly defined.

HOW: The scope of the work must also define the execution mode(s) to be followed.

WHEN: The completion date and any critical intermediate dates must also be part of the contractual agreement.

and last, but not least,

HOW MUCH: Commercial terms and financial arrangements.

There are many different types of contracts that will be discussed briefly in the succeeding paragraphs. However, there are two specific alternatives, applicable to all types, that merit separate mention.

-    **The Agency Agreement.** In this type of contract, the contractor is made the agent of the owner and carries all contract activities in the name of the owner.
-    **The Independent Contractor Agreement.** Under this type of contract, the contractor is engaged to act as an independent entity, fully responsible for all actions taken during the execution of the work until acceptance by the owner.

UNDER THE AGENCY CONTRACT, THE OWNER ASSUMES THE RESPONSIBILITY AND LIABILITY FOR THE ACTIONS OF THE CONTRACTOR. IT SHOULD BE AVOIDED.

6.3   Types of Contracts

Contracts for engineering and construction of a chemical facility can fall under any, or a combination, of the following variations.

By Mode of Selection

Negotiated or Competitive

- If you are a hard-nosed experienced negotiator with a thorough understanding of project costs and are fully versed in the technical aspects involved, a negotiated contract could be a very attractive alternative. Unfortunately, few meet these criteria, and the down-to-earth Project Manager should shy from negotiated contracts and promote competitive situations.

- If time is of utmost importance, a negotiated contract would be the quickest way to get a project rolling, at a price. Conversely, a competitive contract would result in lower costs and a potential schedule delay.

  A WELL-THOUGHT-OUT EXECUTION PLAN MUST ALLOW FOR COMPETITIVE CONTRACTING AND SUBCONTRACTING.

By Breadth of Scope

Single or Multiple

- A single engineering/procurement/construction (EPC) contract consolidates responsibilities and leads to more effective execution. The EPC approach, also referred to as design-build, is a very attractive approach but could be costly and time consuming for major projects, especially if a lump sum price is sought.

- The project scope and execution responsibility can also be split into several contracts to be awarded gradually as the project information and design develop, i.e. engineering, civil work, mechanical work, etc. This approach imposes a heavy load on the owner's Project Manager and is suitable for small projects only. However, it could be, and usually is, used by firms doing EPC work.

By Mode of Reimbursement

Lump Sum or Reimbursable

- When the scope is defined in sufficient detail, the lump sum approach would minimize Owner's risk as well as project schedule, since contractors normally assign their best and most productive personnel to the lump sum work.

Lump sum contracts are more appropriate for construction work, especially when the scope can be divided into discrete parts of homogeneous composition. However, they are not recommended for EPC work because of the potential schedule delay and additional costs to both owners and bidders.

- The unit price contract is a viable alternative to lump sum and permits contracting, and even starting construction work, with minimal engineering. Unit price contracts should be converted to lump sum as soon as possible based on take-offs developed from the final design drawings.

- Reimbursable contracts, also known as time and materials (T&M) or cost-plus, are the least desirable for construction work. However, they could be the only alternative for engineering work when the scope has not been fully defined and/or the owner wants to exercise full control in order to protect proprietary information. They should be tied to a fixed fee to discourage the contractor from incurring unnecessary expenses.

- The guaranteed maximum price (GMP) contract is essentially a reimbursable fixed-fee contract with a price cap that is usually established without the benefit of complete engineering. The Owner approves all expenditures and major decisions and agrees to share the underrun with the contractor based on a mutually agreed upon formula.

All of these types of contracts are normally used in industry and each one offers advantages and disadvantages to both owners and contractors.

Two factors must be weighed against each other when choosing the type of contract:

1. Availability of time and resources required for bidding.
2. The degree of risk the Owner is willing to assume.

For example, a single EPC lump sum contract awarded on competitive basis will place most of the risk on the contractor's shoulders but require a very thorough, lengthy and expensive bid preparation process that could discourage contractors from submitting bids.

On the other end of the spectrum, a reimbursable contract, whether competitive or negotiated, requires minimum bidding effort by both parties but leaves most of the risk on the Owner's shoulders. Between these two extremes, there are several options and combinations which, if discretely managed, could optimize bidding costs and schedule and result in an equitable sharing of risks. The best of all worlds would be to have:

- Independent contractor agreements.
- Competitive bidding.

- Single lump sum EPC contracts.

However, that's not a desirable arrangement for an owner trying to protect a proprietary process. In that case, the recommended approach would be a modified reimbursable EPC contract requesting that the contractor execute construction through competitive lump sum or unit price subcontracts.

The most cost-effective way to execute the construction work is through competitive lump sum contracts. Engineering, on the other hand, should be performed on a fixed-fee reimbursable basis to ascertain that the level of engineering effort and details is sufficient to insure a sound and safe design as well as minimal field questions and extras.

6.4   Contracting Strategy Criteria

General

The contracting strategy is an important element of the Project Execution Plan, affecting cost and schedule, that must be developed early in the project between the Project Manager and the Contract Engineer. It must be reviewed and approved at the proper management level.

The contractor/project size fit must be an overriding consideration in the development of the contracting strategy. Engineering for a large project must be performed by engineering firms with sufficient human and material resources to meet the project needs. This will allow for essential controls and ensure adequate staffing at peak periods. Retaining a small contractor, no matter how good, for a large project is a very dangerous proposition. Without sufficient personnel to cover the project needs, the contractor will inevitably fall behind schedule, resort to temporary help, cut corners, or, worse, a combination of the above.

Large engineering contractors may not accept the ground rules of the small project approach, but even if they do, they will inevitably require more hours than small contractors. This, of course, is an important consideration since engineering is normally done under a cost-reimbursable contract.

Construction is normally handled through competitive lump sum contracts and, experience shows, small local contractors can be more effective than large national contractors providing that the work can be done with their regular permanent staff. Many small contractors get into deep trouble when the work exceeds their capabilities. It is important that the execution strategy for a small project recognizes this fact so that the engineering effort can be geared to the preparation of discrete packages consistent with the capabilities of the local contractors.

The construction of a large project could, in theory, be executed by several small local contractors. This, however, is not practical. Even if enough local contractors were available, the complexity and cost of coordinating them would outweigh the potential benefits.

The Project Manager, together with the Contract Engineer, must decide on the best approach to execute the detailed engineering and construction and develop a contracting strategy to do so. The following criteria outline the main options, and the information required for choosing among them, and discuss the advantages and risks associated with each option.

Engineering

Options
- Do it in-house.
- Contract it.
  1. Single contractor.
  2. Multiple contractors.
- Do part in-house, part by contractor.

Required Information
- Design expertise required.
- Estimated hours by discipline.
- Availability of specialized contractors close to the site.
- Availability of CED and division personnel.

Discussion

- Doing all engineering in-house is not a real alternative. The Owner should not assume the responsibility and liabilities associated with structural design even if it had qualified professional engineers. That responsibility should always be delegated to a well-qualified engineering firm.
- Using a single engineering firm would simplify coordination and consolidate responsibility for potential delays. However, it should be made very clear that project management and control will be the Owner's responsibility.
- Using small local contractors familiar with the plant, for structural and/or electrical detail design could be less expensive and also promote local goodwill. However, they should be screened carefully to make sure they have adequate resources.
- In the mixed approach, the Owner could do the instrumentation and maybe the purchasing. The coordination of all design activity, in-house and contracted, is the responsibility of the Project Manager.

Construction

Options
- Single contractor.
- Multiple contractors.

Required Information
- Construction hours.
- Schedule requirements.
- Local contractors' size, availability and experience.

Discussion

- Retaining a single contractor would simplify the construction management effort but may be more expensive since a single contractor would probably need to subcontract some of the work and incur another level of construction management costs. It that's not the case, then a single contractor could be the right choice, provided it does not impact the schedule.
- The use of multiple contractors would be beneficial to the schedule since construction work would start well before engineering is completed.

6.5   Selecting EPC Contractors

A small project, by definition, could not be executed under an EPC contract without ceasing to be a small project. However, it could very well occur that what was originally conceived as a small project grows beyond its parameters and the Project Manager is suddenly handling a conventional project requiring a different contracting approach. Discussing the selection of EPC contractors then seems appropriate. The discussion addresses the recommended reimbursable EPC contract with construction through competitive lump sum subcontracts.

Although the evaluation and selection process is essentially the same for all contracts, an EPC contract is more complex and requires more work and attention to details. The basic steps in the contractor selection process are:
- Preparation of bid package.
- Bidders selection.
- Preparation of bids.
- Bids evaluation and contractor selection.

Preparation of Bid Package

General

A thorough bid package insures a meaningful bid. The bid packages required for lump sum EPC contracts for large facilities are elaborate, voluminous and require a lot of attention to details. On the other had, the bid package for a reimbursable contract, no matter how large, could be quite simple but still requires attention to all details. If the scope is not accurate and the instructions precise and clear, the bids will not be comparable and their evaluation would be, at best, cumbersome.

A bid package must address commercial terms as well as technical details. The preparation of the commercial selection is the responsibility of the Contracts Engineer. The Project Manager assembles all the technical details and is responsible not only for the technical section but for the overall completeness of the package. After all, it is the Project Manager who has to live with the contractor and make it perform.

Commercial Section

The commercial section must contain the following information:

- Instructions to Bidders

    - Bid submission date.
    - Award date.
    - Number of copies required.
    - Addresses/telephone numbers.
    - Handling of questions/answers.

- Contractual Conditions

    - List of acceptable reimbursable costs.
    - Type of contract.
    - Typical contract.
    - Secrecy requirements.

- Proposal Contents

    - Schedule of reimbursables.

- Bare reimbursable salaries.
- Schedule of overheads (home office and field).
- Travel policy.
- Reproduction costs.
- Proposed execution approach.
- Relevant experience.
- Key personnel resumes.
- Work backlog.
- Estimated home office hours broken down by disciplines.
- Proposed project schedule.
- Proposed construction subcontractors.

Technical Section

The technical section is essentially the scope of the work. It must provide a good understanding of the technical aspects of the project so that the bidders can prepare meaningful proposals including accurate estimates of the engineering and construction management hours.

Process Information

- Type of process: Solid/liquid/gas, batch/continuous, pressure, temperature, organics/ inorganics, unit operations, retrofit, hazards, etc.
- Process background: Known process/prototype, stage of process development, scale-up factor, etc.

Hardware Information

- Equipment count by type and size corrected for growth allowance.
- Materials of construction.
- Electrical: Motor count, average size, substation and distribution system requirements, uninterruptable power supply requirements, etc.
- Instrumentation: Control philosophy, sophistication control room requirements, etc.

- Buildings and structures: Number, dimensions, materials, function, etc.

Site Information

- Location.
- Grass roots/existing site.
- Availability of utilities.
- Soil conditions.
- Construction, layout areas.
- Site access, parking.
- Environmental information.

Execution Information

- General: New plant/rehab/retrofit, shutdown work, plant rules, special control requirements, completion date, long delivery items, etc.
- Design: General specifications and standards, layout criteria, expected life, etc.
- Construction: Direct hire/subcontracted, union/merit shop, etc.

Contractors' Work

- Phase 0/Phase I: Process responsibility, support owner, extent and areas of support location, etc.
- Phase II: Limits of responsibility, owner's approvals, design manuals, mechanical catalogs, reporting, control requirements, etc.

Owner's Supply

- Process design.
- Site information.
- Permits.
- Etc.

Bidders Selection

Potential bidders must be screened thoroughly to determine their technical competence as well as their organizational capability and/or willingness to meet project specific requirements, size fit, construction approach, etc. Contractors should not be placed on the bidders slate unless they are considered fully capable of performing quality work within the criteria established in the project execution plan. The criteria are the responsibility of the Project Manager, who must consider, as a minimum, the following factors:

- Relevant Process Experience

  - Organic/inorganic/salts.
  - Developmental/First of a kind.
  - Specialized process design programs.

- Relevant Execution Experience

  - Experience in geographical area.
  - Retrofit/rehab work.
  - Remote areas, construction camps.
  - Direct hire/subcontracting, construction management.
  - Control of cost, schedule; quality.

- Specific Engineering Expertise

  - High-pressure-temperature technology.
  - Materials handling.
  - Special equipment applications.
  - New technology applications.

- Size Fit

  - Engineering staff: 3-6 times expected peak.
  - Average size of projects normally handled.

- Location of Engineering Office

- Interest Shown by Management

- Work Load

Preparation of Bids

The Owner's direct participation during this phase is limited to answering requests for clarifications and/or additional information. All of these should be handled officially through the Contracts Engineer, who in turn must refer technical matters to the Project Manager.

All answers must be documented in a formal manner with copy to all bidders unless they refer to a particular bidder's proprietary know-how or alternate proposal to execute the project in a uniquely different manner. These unique alternatives could be welcomed, and even encouraged, but should never be considered unless the bidder responds first to the bid package.

Bids Evaluation and Contractor Selection

Bids evaluation must be a well-organized and deliberate effort that culminates in the selection of a technically qualified contractor capable of executing the work in the most cost-effective manner.

It is a good practice for the Project Manager and Contract Engineer to conduct independent technical and commercial evaluations with the technical evaluation being conducted without any prior knowledge of the proposed commercial terms. It is also a good practice that technical evaluation criteria be prepared by the Project Manager with input from process engineering and plant operating groups.

Technical Evaluation

The following is a checklist of the points that should be considered for technical evaluation criteria. The relative weight applied to each will depend on project specific conditions.

General Considerations

- Approach to project execution.
- Approach to construction.
- Proposed control system.
- Management commitment to project.

- Flexibility.
- Approach to procurement.
- Size of project handled.
- Size of company.
- Stability of company.

Specific to Project

- Process:
  1. Relevant experience of proposed team.
  2. Availability of up-to-date design systems and tools.
- Design disciplines:
  1. Relevant experience.
  2. Depth and breadth.
- Key personnel:
  1. Experience together in function.
  2. Project Manager/Engineer.
  3. Process Leader.
  4. Cost Engineer.
- Understanding of scope of work.
- Experience in geographical area.
- Experience with other Owner's projects.

Technical evaluation is a subjective process and, as such, should reflect the consensus of several evaluators working both independently and as a group. Appendix E is an example of an actual technical evaluation for a conventional project.

Commercial Evaluation

The commercial evaluation is essentially an arithmetical exercise to determine the optimum combination of estimated hours, bare salaries and contractor's mark-ups. Bidders must be required to submit this information as part of the bid.

- Some bidders will do their homework and prepare realistic hours estimates. Others may try to impress the Owner by submitting the lowest estimate they can ethically justify. The Owner's best protection against this eventuality is to prepare in-house estimates and apply them to all the bids. However, some contractors can perform the same work in fewer hours than others, either by running a lean and mean organization or by being very proficient with up-to-date drafting techniques, such as 3D CAD. If that difference is detected during the evaluation, it should be taken into account in the final decision.

- Bidders usually submit their salary schedule as a range for each discipline; some will also indicate what the average is. The average has no contractual value and could be misleading. The most objective evaluation approach would be to calculate the overall average rate for each bidder and apply it to the total hours estimated by the Owner. The overall average rates must be calculated using the middle of the ranges and the discipline mix proposed by each bidder.

- The contractor's mark-ups represent the total of payroll added costs (PAC's), overheads and profits.

  The PAC's are actual out-of-pocket costs, dictated mostly by laws and regulations, and could vary from area to area to reflect local regulations. The bidders must provide, at Owner's request, a verifiable breakdown of these costs.

  The overheads and profit reflect contractor's cost of doing business, operating costs, management, marketing, etc., and is usually the determining factor in the commercial evaluation.

Another important consideration in the commercial evaluation is the bidder's reaction to the Owner's pro forma contract included in the bid package. A strongly negative reaction could be an insurmountable obstacle, especially if it pertains to ingrained management principles. However, this would be an extreme case and in real life reasonable people will negotiate a mutually satisfactory agreement.

Contract Award

Occasionally the best technical rating coincides with the most favorable commercial conditions. The decision then is very easy. Unfortunately this is not a frequent occurrence and a choice must be made after evaluating the extent of the differences and the potential impact on the overall project performance, quality, schedule and cost. The final decision should be based on the following considerations:

- All the bidders were prequalified and found capable of a good performance, and unless an unexpected factor turned up in the final evaluation, they are still capable.

- Usually many of the technical reservations are related to the proposed key personnel and can be easily corrected by requesting alternative candidates and securing formal commitments for the preferred ones.
- The final contractor's costs on a reimbursable contract depend on total hours, actual salaries, and mark-ups. The latter is the only contractually guaranteed factor. The total hours would be a reflection on the contractor's management and technical ability and the actual salaries could fall anywhere within the ranges indicated in the proposals.

Some Owners engage in the practice of bid shopping, using information or ideas from one bidder to squeeze concessions from another. This practice not only tests the limits of ethical behavior but will ultimately backfire on the user. Eventually bidders will catch on to this practice and include a negotiation factor in their bids and the Owner will likely end up paying a higher price.

6.6   Subcontracting Construction Work

Overview

By definition, the small project Manager acts as general contractor and, as such, is also the construction manager, with an active participation in subcontracting, working closely with the Contract Engineer. As already mentioned, the most effective way to perform the construction work is through competitive lump sum contracts. This section is specifically addressed to that type of subcontracting.

The subcontracting process entails the same basic activities of EPC contracting:

- Bid Package.
- Bidders' qualifications.
- Bidding.
- Bid analysis and contract award.

The Project Manager and the Contract Engineer must have a good understanding of the scope and the cost of the work in order to evaluate bidders and analyze the bids. To do so, they need a good cost estimate with detailed or semi-detailed take-offs.

Ideally the original cost estimate should contain sufficient details to recast the various costs associated with each particular subcontract. However, if the details are not available and/or the scope of the work has changed, it is up to the Project Manager to provide a cost estimate with sufficient details.

Bid Package

General

The bid package must provide a clear understanding of the scope of work, execution requirements and commercial conditions that will permit the bidders to prepare a realistic lump sum proposal. It must include technical, and commercial information as well as clear project execution and bidding instruction.

Technical Information

This section is usually prepared, under the supervision of the Project Manager, by whomever performs the detailed engineering. It must include all pertinent drawings and specifications.

Project Execution Instructions

The Project Manager must have a direct participation in the preparation of the execution instruction. They must address:

- Schedule requirements.
- Testing extent and responsibility.
- Site conditions affecting cost and schedule:
  1. Soil data.
  2. Plant rules.
  3. Work interferences.
  4. Availability of construction utilities.
- Reporting requirements.

Commercial Information

The commercial section of the bid package is the responsibility of the Contracts Engineer and must address the contract, invoicing procedures, handling or extras, etc.

Bidding Instructions

These require the direct participation of both the Project Manager and the Contracts Engineer. In addition to information concerning bid submittal date, format and required pricing structure, the bidding instructions must include a complete list of the information that must be provided by the bidders as a prerequisite to having their bids evaluated, including:

- Cost breakdown by materials, labor and indirects.
- Materials take-offs and estimated work hours by craft.
- Unit prices and work hours to serve as the basis to evaluate extra work.
- List of proposed subcontractors, if any.
- Preliminary schedule and execution plan showing proposed staffing.
- Key supervisors' resumes.
- Exceptions to proposed contractual conditions, if any.

The Project Manager must avoid imposing project control requirements that could be beyond the capability of the small subcontractors usually employed in small projects. A better approach is to wait and see what the bidders propose.

If the proposed controls are not satisfactory but the bidder is still preferred by other overriding reasons, the Owner must make provisions to complement the contractor's capabilities and adjust the projected cost accordingly.

Bidders' Qualification

The Project Manager must work with the Contract Engineer to qualify and select capable bidders. If the project is in an existing plant site, the selection process must be done in close cooperation with the plant personnel since they must approve all contractors working in their plant. In the case of small projects, especially if the execution approach calls for multiple subcontractors, the volume of work will not be large enough to attract national contractors and the choice will be among small or medium-sized, local contractors.

Usually when the work is to be done in an existing plant, the potential bidders are well known to the local personnel and may even be working at the site on a regular basis. This, or course, will expedite the bidders' qualification process. However, it could also be a source of potential trouble:

- Pressure to favor a particular contractor.
- Reluctance to develop new bidders for plant work.
- Having the same contractor doing lump sum subcontract work and reimbursable plant work.

In that case, the Project Manager must walk a very fine line when trying to qualify new contractors to bid against the regulars.

When the project site is a new location and the information available on local contractors is limited, the bidders' qualification process must be more elaborate. Contractors must be interviewed and their references must be checked to determine their qualifications for the specific project.

Size

  - Volume of work - lump sum/reimbursable.
  - Average size of project.
  - Largest project staffing.
  - Permanent employees.

Relevant Experience

  - With similar projects.
  - In area.
  - In specific plant.

References

  - *Dun & Bradstreet.*
  - Banks.
  - Safety record/insurance rating.
  - Other clients.

Bidding

Once the bidders have been qualified and accepted by the plant, the bid packages are given to them and a pre-bid meeting is held with all bidders, preferably about one week later. The pre-bid meeting is co-hosted by the Contracts Engineer and the Project Manager. The objective is to:

  - Insure good understanding of scope and contractual conditions by every bidder.
  - Answer and clarify questions for the benefit of all bidders.
  - Make any last minute changes to the scope and/or contractual conditions.

The Contract Engineer will issue minutes of the pre-bid meeting within two days.

All questions raised by the bidders during the bidding period will be answered formally with copies of all answers sent to all bidders, except for items deemed to be proprietary by the bidder. Technical questions will be handled by the Project Manager and contractual questions by the Contract Engineer.

Bid Analysis and Contract Award

The Project Manager and the Contract Engineer will conduct independent reviews of the technical and commercial parts, respectively.

As part of the technical analysis, the Project Manager must analyze bidder take-offs to ascertain the bidder's understanding of the scope and will also analyze unit costs and estimated hours for reasonableness and consistency. To do so, the Project Manager must develop, using the same information sent to the bidders, independent take-offs and estimated work hours. Major deviations from the estimates must be discussed with the Contract Engineer to determine how to approach the bidder on the subject.

If the low bidder has reasonable take-offs and unit prices and shows good understanding of the scope, the decision will be very easy. Unfortunately, that's not always the case.

In most cases, final meetings have to be held with each of the two most promising bidders to clarify questions, ascertain their understanding of the scope and confirm the validity of their bids. Usually, as a result of these meetings, the bidders have to make changes and/or additions to the bids.

Both the Project Manager and the Contract Engineer must be very careful that their statements won't relieve the bidder of any of their obligations nor give anyone an unfair advantage over the other. Section 6.8 Do's and Don'ts of Contracting, offers more advice on the subject.

Finally, the Project Manager and Contract Engineer will prepare a formal bid analysis with their joint recommendation and have it reviewed and approved at the proper management level.

6.7   Contracting Engineering Services

The extent and complexity of the engineering services associated with small projects is substantially less than for major projects. For small projects, engineering firms are usually retained to do only the detailed engineering and assemble technical specifications to purchase equipment and materials and subcontract construction work. Occasionally, the design work is broken down by disciplines and farmed out to small, local consulting firms or even done partly in-house. Purchasing and expediting may or may not be included in the detailed engineering contract depending on the availability of Owner personnel to assume that responsibility.

As mentioned in Section 6.1 Overview, engineering should be contracted on a reimbursable basis to allow the Project Manager to exercise close control of quality and overall project cost. Engineering firms doing design on a lump sum basis may try to minimize the number of hours used by reducing the amount of documentation and detail shown in the drawings and specifications. They will argue that doing so does not affect the soundness of the design. In many cases, that will be true; however, the number of drawings and the detail of information shown on them may not be sufficient to prepare a meaningful construction bid nor to build the project properly. The result would be higher construction costs for:

-   Field extras.
-   Additional field supervision and engineering follow-up.
-   Costly field errors.

Frequently, all the detailed engineering, and even purchasing, is contracted with a single engineering firm. In that case, it becomes very important that the contractor understands that the scope of the contract work is strictly detailed engineering and procurement and that the Owner will assume the project management and control responsibility.

Occasionally, the Project Manager will run into a proposal to do the detail engineering on a lump sum basis at a fraction of the budget cost. Before falling into the temptation and accepting the offer, the Project Manager should first:

•   Ask the bidder for a complete list and scope of the documentation intended under the lump sum.
•   Sit down with the operating personnel and CED specialists to determine what is the minimum information required for proper record keeping and plant maintenance.
•   Make sure that the bidder understands the bid requirements and has included the related costs in the proposal.
•   Reaffirm the Owner's right of review and approval, establish a review and approval cycle and make sure that the resulting schedule and cost impact is included in the lump sum.
•   Ask the bidder for a formal confirmation of all of the above and, if required, a revised proposal.

Although contracting engineering services for small projects does not entail the use of the more sophisticated procedures followed in large projects, the considerations are essentially the same. The contracting guidelines in Section 6.5 Selecting EPC Contractors, are intended for conventional projects, but they could also be very useful for small projects.

6.8 Do's and Don'ts of Contracting

The following bits of wisdom have been voiced by experienced contract engineers and project managers:

- DO have a scope and an execution plan before starting the contracting process.
- DO review the commercial bidding instructions before they are issued to insure that the special conditions documents are consistent with the particular project.
- DO insist that bidders respond to the base case before even considering their proposed alternatives.
- DO develop your own estimates of the work involved prior to receiving the bids.
- DO review bidders' take-offs and challenge them if they are more than 20% off your own take-offs.

BUT

- DON'T let them know whether they are high or low.
- DON'T ask for price adjustments without a corresponding change in the scope.
- DON'T use general terms like "smooth," "exact level," etc. Give tolerance figures.
- DON'T accept any price reduction during the bid evaluation without a real justification.
- DON'T change technical specifications during the negotiations without checking with the appropriate specialists.
- DON'T socialize with bidders during the bidding process.
- DON'T award a lump sum contract with less that 90% engineering.
- DON'T award a contract until all questions have been satisfactorily answered in writing by the bidder.

AND

- NEVER use a purchase order to award contracts.

6.9   Typical Contracts

The contract is the fundamental document in project execution, as mentioned in Section 6.2. It defines the what, the who, the how, the when and the how much. It includes a basic document (the agreement) and a series of attachments (sometimes called appendices or schedules) that define all details of the scope, regulates the communications between contractor and Owner and establishes the rights and obligations of each.

The Agreement

The agreement is the document that recapitulates, by either direct mention or reference, all the elements of the contract (scope, schedule, general and special conditions, commercial conditions, etc.) and formalizes them through the signatures of the parties involved. The agreement can also be the vehicle to clarify, reinforce and even override some of the conditions incorporated in the attachments.

The following information should be included in the agreement:

- Clear identification of all signing parties.
- Brief description and exact location of the work.
- Project completion date.
- Preset fixed costs:
  1. Lump sum.
  2. Fixed fees.
  3. Guaranteed maximum price.
- Liquidated damages clause, if any.
- Arbitration procedure.
- List of all attached contractual documents with clearly established precedences.

Scope of Work

The scope of work documents must include:

- Scope on which the bid is based.
- Minutes of pre-bid meetings.
- Addenda and/or clarification of the scope issued during the bidding process.

- Minutes of the pre-award meeting.

On a lump sum EPC contract, the scope documentation is extensive and bulky. It includes:

- Drawings.
- Engineering standards.
- General specifications:
  1. Civil.
  2. Mechanical.
  3. Electrical.
- Detailed specifications:
  1. Equipment.
  2. Site.
- Execution instructions:
  1. Owner's approval.
  2. Documentation requirements.

On the other hand, the scope of a reimbursable contract requires minimum documentation; it could be as simple as a one-page, verbal description.

General Terms and Conditions

The general terms and conditions set the rules that regulate the implementation of the contract and define the contractual rights and obligations of each party. The most relevant, typical clauses are summarized below:

- The contractor operates as an independent contractor.
- The contractor shall comply with all federal, state and local laws and codes including all provisions of Executive Order 11246 concerning equal opportunity employment.
- The contractor shall, if so required, remove from the job any employee the owner deems to be incompetent or undesirable.
- The site and subsurface information provided by the owner is supplied for the contractor's convenience only. The contractor shall confirm it to its satisfaction, assume the responsibility and promptly notify the owner in writing of any actual or latent physical condition differing from those indicated in the contract.
- The contractor holds the Owner harmless from any claims and associated costs related to infringement of any patent arising from the use of materials

supplied by the contractor and/or personal injury and/or property damage resulting from work performed by the contractor.

- The contractor shall be responsible for the protection of all work in progress and shall comply with all safety regulations and standards.

- The Owner has the right to inspect at all times any work covered by the contract on-site and off-site.

- The contractor shall maintain a clean environment and avoid any kind of pollution.

- The Owner reserves the right to direct the contractor to schedule the order of performance of its work in a manner which does not unreasonably interfere with the performance of the work.

- The contractor shall submit advance working schedules and take the necessary steps to achieve them. Failure to do so provides ground to terminate the contract.

- The Owner may at any time make changes and/or request the contractor to perform extra work. The contractor shall comply immediately, if so requested and submit a formal estimate of the cost and schedule impact within 10 calendar days.

- If the contractual schedule is delayed for any reason beyond its control, the contractor shall give the Owner written notice at the time that delay occurs. Failure to do so shall be sufficient reason for denial of all schedule extension.

- If the contractor fails or refuses to provide sufficient resources to perform the work, the Owner has the right to terminate the contract for default after giving the contractor prior written notice and has then the right to do whatever is required, including taking over the contractor's resources, to complete the work.

- The Owner also has the right to terminate the contract at any time for its convenience and reimburse the contractor for all reasonable expenses incurred to date.

- Any controversy or claim arising out of the contract shall be settled by arbitration in accordance with the rules of the American Arbitration Association.

The following subjects should also be addressed in the general terms and conditions:

- Insurance requirements.
- Requirements for contractor safety program.

- Handling and approval of contractor's drawings, purchase orders, subcontracts and materials certification.
- Requirement for contractor's contingency plan.
- Owner's right to examine contractor's records.
- Contractor's reporting requirements and progress and coordination meetings.
- Employment eligibility verification.

Special Terms and Conditions

The special terms and conditions are intended to address matters specific to each project and/or each site such as:

- Availability of utilities.
- Services provided by Owner.
- Project schedule.
- Plant regulations.

Proposal Information

Some of the documents submitted with the proposals must be included in, and officially made part of, the contract. Those more frequently included are:

- Pro-forma proposal offer.
- Schedule of prices broken down as requested.
- Key personnel list.
- Contractor's safety and CICE programs.
- Schedules.
- Safety data forms.

Reimbursable Costs Schedule

This schedule is applicable only in cost reimbursable contracts and must clearly identify the cost items that will be reimbursed to the contractor under the terms of the contract. It would be to the Owner's advantage to include the following clauses:

- Make all clerical and secretarial help non-reimbursable and part of the overhead.
- Make the cost of word processing, CAD equipment and the like part of the overhead.
- Don't accept mark-ups or overhead charges on agency personnel (jobshoppers), local field hires or subcontracts. Treat them as a pass-through charge.
- Don't pay overhead on overtime work.

CHAPTER 7
DETAILED ENGINEERING

7.1 Overview

Detailed engineering is the project phase where the difference in execution between small and major projects is most noticeable. It is also a phase that presents opportunities of substantial cost reduction without jeopardizing quality. In the context of this guideline, detailed engineering includes:

- Upgrading P&ID's to approved for design (AFD) status.
- Preparation of detailed project specifications - approved for design.
- Execution of the detailed engineering and preparation of approved for construction (AFC) drawings for all disciplines.
- Preparation of purchasing packages for competitive bidding of construction work.

On a major project, the Owner would normally select a full service engineering and/or construction firm to act as general contractor under an EPC contract to perform all the functions required to carry the project to a successful completion. These functions normally comprise much more than detailed engineering, and include:

- General specifications.
- Estimating.
- Scheduling.
- Cost Control.
- Procurement/subcontracting.

- Project management.
- Construction or construction management.

To perform all of these functions effectively, the general contractor must have a well-structured organization with formal procedures and lines of communication. Such an organization entails a high overhead burden:

- The ratio of supervisory and support personnel to designers is high.
- Reviews and approvals are elaborate.
- Response is slow.
- Learning curve is long.
- The relative cost is higher.

However, the degree of complexity of a major project requires a complex organization in order to insure quality, schedule and cost control. In a major project, a general contractor is a must, and the price has to be paid to insure the integrity of the project.

The lesser degree of complexity of a small project does not warrant the complex and formal organization provided by an EPC contractor. Experience shows that the ratio of unproductive to productive hours increases as the size of the project decreases. Substantial savings can be achieved through the small project approach where the Owner assumes the project execution and control responsibilities and uses engineering and construction contractors, as well as in-house resources, to perform detailed engineering, procurement and construction. In doing so, the Owner becomes, in fact, the general contractor.

To be successful, the small project approach must be followed only:

- On the right size project.
- If the Owner has adequate human resources with the right experience at the right time.

Following the small project approach is not a decision that can be taken in haste. It requires careful consideration, participation and approval at the right management level.

The purpose of this chapter is to:

- Give the Owner's Project Manager an insight into the workings of an engineering office with its execution and checking procedures, internal documentation, etc.

- Discuss the various ways in which detailed engineering could be executed to minimize project costs without impairing quality and alert the Project Manager of the potential pitfalls.
- Provide guidelines for the coordination of the detailed design activities to fill the vacuum left by the general contractor.

The discussions are limited here to only the execution of detailed engineering. All other related functions, project and construction management, purchasing and subcontracting as well as project controls are discussed in other chapters.

7.2 Execution by Contractor

An Owner's Project Manager assigned to the office of a sophisticated engineering contractor is seldom aware of all the behind-the-scenes activities that must take place before the final product is submitted for Owner's approval.

All of those activities are regulated by complex internal procedures and protocols, such as:

- Scope definition and change control.
- Checking of design.
- Reviews and approval at various levels.
- Intersquad coordination.
- Progress evaluations.
- Resource allocation.
- Variance reports.
- Cost tracking.
- Design procedures.
- Communications formats.
- Others.

In the case of large contractors, these procedures can be very strict and inflexible.

Some of those procedures are intended only to protect the contractor's interests, control internal costs and gather data for internal use. Most of them would not apply to the small project approach.

Other procedures and activities are directed to controlling project quality, cost and schedule. They must be included, at least in essence, in the protocol of all projects.

Finally, and especially in larger organizations, there may exist procedures and activities that serve no useful purpose other than to impress the client. The Project

Manager should review these carefully and, in the case of reimbursable contracts, take exception and refuse to pay for them.

Although this is not always the case, smaller contractors are more flexible and informal than large ones.

This section provides insight into the behind-the-scenes workings of an engineering contractor's office, discusses the procedures that govern them and focuses attention on those that should be followed and/or enforced by the Project Manager during the detailed engineering phase.

Basic Engineering

Immediately after receiving a package from a client, the first thing contractors do is review it for completeness by going through their standard questionnaires to determine what is missing and how much basic engineering must be done to bring the project to the point where detailed design can start and proceed efficiently. This is usually much more than most owners think.

To illustrate the point, it should suffice to note that the Phase I package described in Chapter 3 requires approximately 10-12% of the total home office man-hours, while most contractors estimate that their basic engineering consumes approximately 20% to 35% including process design.

The following paragraphs list the main activities that must be accomplished before the project can be declared approved for design.

NOTE: Items marked with an asterisk require direct participation of the Owner's Project Manager.

- Review client's process design and material balances.
* Prepare and issue project coordination procedure and assign responsibilities
* Develop engineering man-hours budget for each discipline.
* Prepare project execution plan, determining:
  1. Firm engineering schedule.
  2. Client's review and approval requirements.
  3. Preliminary construction schedule.
  4. Subcontracting strategy.
- Refine P&ID's for approved-for-design issue:
  1. Confirm hydraulic calculations (line sizing) for compressor circuits and establish compressor differential.
  2. Confirm relief protection philosophy, select location of major relief valves and determine vessel and exchanger design pressure and temperatures.

    3. Confirm hydraulic calculations on all pump circuits and establish pump NPSH and minimum equipment elevations.

    4. Confirm and, if required, complete sizing of lines and control valves.

    5. Confirm that all relief valves as well as vent and drains are shown.

- Review all equipment datasheets and incorporate all pertinent mechanical information to make them ready for procurement.
- Prepare project control documentation, including:
  1. Drawings and requisitions lists.
  2. Vendors lists.
  3. Line schedules.
  4. Painting and insulation schedules.
  5. Procurement logs.
  6. Equipment and instrument control indexes.
* Compile and issue project specifications.
- Obtain site-related data if not available at contract award, such as soil conditions, site surveys, topographical maps, etc.
- Prepare equipment arrangement drawings.
- Develop electrical one-line diagrams and electrical loads lists.

Detailed Design

The detailed design phase or production engineering, as some people call it, consumes approximately 50% of all home office man-hours. It is during that phase that the bulk of the documents directly required for procurement and construction are generated:

Civil Engineering

- Site studies, plot plans.
- Soil studies, piling design.
- Storm and process sewers.
- Roads, fencing, parking, paving.
- Structural calculations - concrete, steel.
- Foundation drawings, rebar schedule.
- Structural steel drawings - plans, elevations.
- Miscellaneous building design and drawings.

Piping Engineering

- Equipment arrangement drawings.
- Piping plan and elevation drawings.
- Isometric drawings.
- Model.
- Stress analysis.
- Bills of materials and requisitions.
- Valve schedules.
- Heat tracing system layout.

Instrument Engineering

- Instrument specifications and data sheets.
- Material list and hook-up drawings.
- Interlock system.
- Drawing for emergency electrical system.
- Standard drawings for field instruments.
- Layout drawings for cable ducts, joint boxes and joint plates.
- Schematic drawing for instrument air supply piping.
- Schematic drawing for instrument heat tracing.
- Board face layout and wiring.
- Wiring diagrams for interlocks.
- Requisitions suitable for purchasing of all instrumentation items.
- Review bids for conformance to specifications.
- Uninterrupted power supply (UPS) power source and distribution.

Electrical Engineering

- Mechanical and material specifications for all electrical services.
- Single line wiring diagram, motor lists.
- Load schedule, power factor correction scheme, relay coordination study and diagram.

- Lighting system drawings showing arrangement of lighting panels, lighting requirements and specific details as required.
- Layout of power cables and specific requirements for switchgear and motor control centers, including mounting details.
- Grounding drawings and specific details.
- Emergency lighting source for control room and field.
- Layout and detail drawings of cable duct, if any.
- Arrangement of above-ground conduit and cable trays.
- Layout drawings of overhead piping for power, lighting and instruments.
- Layout of substation room including arrangement of switchgear, MCC and outdoor transformers.
- Outline of MCC and switchgear.
- High-voltage and low-voltage switchgear schedules.
- Motor control diagrams.
- Wiring diagram of switchgear.
- Hazardous area classification drawing.
- Requisitions suitable for purchasing of electrical items.
- Composite bill of materials.
- Emergency power source.

Insulation Painting and Fireproofing

- Schedule of type, thickness and covering of insulation of equipment and piping.
- Requisitions to purchase insulation and painting materials and services
- Fireproofing schedule.
- Drawings covering specific details for fireproofing.

Coordination and Control

Even a small project ($3 to $5 million) will require a staff of 15 to 25 people during the peak of the detailed engineering effort. Each of them must receive the correct information at the right time and, in turn, complete the assigned tasks correctly and on time. Additionally, all of the activities must be performed within the frame of the budgeted hours. This can only be achieved by imposing a project control system. A control system, regardless of the size of the project, must include:

- Detailed project schedule.
- Coordination.
- Quality assurance and control (QA/QC).
- Productivity and progress measurements.

For large projects, where the home office staff peaks in the hundreds, the system must be elaborate and expensive but absolutely necessary. On a small project, all of the required control functions could be performed by an experienced project engineer with assistance from the discipline leaders. However, a large contractor used to handling megaprojects might find it impossible to adapt to the realities of small projects. As mentioned in Chapter 10, it is up to the Project Manager to analyze the organization of potential contractors and steer away from those with large and inflexible control organizations.

7.3   Small Project Execution Options

General Considerations

As mentioned before, on a small project, the Owner acts as a General Contractor assuming the responsibilities of:

- Process engineering.
- Project management.
- Construction management.
- Subcontracting.
- Project controls - estimating, scheduling and cost control.

Occasionally the Owner would also perform:

- Procurement.
- Instrumentation engineering.

Rarely, only on very small projects, the Owner could also perform some mechanical, electrical and/or piping design. Civil and structural engineering and, usually, mechanical, electrical and piping design are assigned to one or more engineering firms, preferably small or medium sized ones.

The execution of detailed engineering on a small project is a mixed bag. It could be performed by any combination of:

- In-house engineering - CED, division or plant.
- Engineering firm - large or small.

The selection of the proper mix is a pivotal decision that has a lasting effect on the fate of the project and requires participation of the Contract Engineer as well as participation and approval of management. It requires very careful consideration of:

- Range and depth of engineering expertise required.
- Availability of in-house personnel.
- Size of project and engineering hours.
- Size and qualification of proposed engineering firms.

It is up to the Project Manager to develop sufficient data so that the correct decision can be made. The Project Manager must:

- Estimate the hours required to perform the detailed engineering in-house as well as by contractors.
- Work with the specialists to determine if any particular engineering expertise is required to insure safety and health protection or for special technology needs.
- Evaluate the capabilities of the in-house resources and their availability on a timely basis consistent with the project schedule.
- Secure commitments at the proper management level.
- Evaluate, together with the Contract Engineer, the availability and capability of local engineering firms.

In-House Engineering

In-house engineering is probably the least expensive approach of any case, particularly when it can be performed by personnel that would be involved in the monitoring of a contractor's work anyway, such as CED instrument engineers.

In small projects, CED and/or plant mechanical and electrical engineers can take hands-on participation in equipment design and purchasing specifications.

Some plants have engineering departments with design and drafting capabilities and are quite capable of doing equipment layouts and piping design as well as electrical and instrumentation schematics and loop drawings.

A WORD OF CAUTION:

> DOING ALL DETAILED ENGINEERING IN-HOUSE IS NOT A DESIRABLE ALTERNATIVE. THE OWNER SHOULD NOT ASSUME THE RESPONSIBILITY AND LIABILITIES ASSOCIATED WITH STRUCTURAL DESIGN EVEN IF QUALIFIED PROFESSIONAL ENGINEERS ARE AVAILABLE. THE RESPONSIBILITY FOR CIVIL AND STRUCTURAL DESIGN SHOULD ALWAYS BE ASSIGNED TO WELL-QUALIFIED ENGINEERING FIRMS.

For a small project, executed at plant level by plant project engineers, in-house engineering is the ideal approach provided that adequate resources are available in the plant engineering department, because:

- The required engineering design hours could be as low as 50% of those required by an outside contractor.
- The out-of-pocket costs would be minimal.

Small projects, wherein most of the detailed engineering can be done by the plant engineering department, should also be managed at the plant level by a plant project engineer, rather than by CED. Communications, coordination and supervision should be more effective since the project engineers, design engineers, draftsmen and even the "client" will be working, if not for the same supervisor, at least in the same operations unit within the Owner's organization.

Contracted Engineering

When the required engineering effort exceeds the resources of the local plant engineering department, the work must be performed by a contractor.
There are several options to contract engineering work:

- To a single large engineering firm.
- To several small engineering firms.
- To a single small or medium sized engineering firm.

The large engineering firm option will be the most expensive and should be avoided unless the project demands a very specific expertise only available through a large firm.

Most large contractors, with their elaborate communication and checking procedures, are inflexible and could not function without them. Forcing them to do so would be flirting with disaster. The potential savings of the small project approach would be limited to the hours related to project control function retained by the Owner. No savings should be expected in project management since a large contractor would assign a project manager or project engineer to the job anyway.

When a special expertise is required, it may be possible to avoid the large engineering firm by shopping around for individual experts to support the efforts of smaller contractors.

Another option is to farm out the detailed engineering among several specialized local firms, familiar with the plant and its requirements. This approach would be ideal for a retrofit job where a local civil engineer can design a couple of foundations and the engineering section of the friendly local electrical contractor could handle a few motors connected to an existing MCC, and so forth. This option, although financially attractive, could easily overload the Project Manager to the point where the entire project would suffer.

Retaining a small or medium sized full-service engineering firm to perform all the detailed engineering, and even the purchasing, would be the ideal option for the run-of-the-mill project that does not require any particular design expertise.

Small contractors usually have and, in this case, the selected contractor must have:

- Adaptability to accept the direct participation of the Owner's Project Manager in the design activities.
- Qualified personnel with the flexibility to perform without their normal organization constraints.

Should the Owner fall short in providing the Project Manager with staff support to perform any of the design-related functions under its responsibility, a full-service contractor could provide qualified back-up on minimal notice.

7.4 The Project Manager as General Contractor

In the absence of a general contractor, the Project Manager must step in to fill the breach and insure, through hands-on action and/or direct supervision, that the essential activities required for proper coordination and control of the detailed engineering phase are performed on a timely basis. Those activites include:

- Engineering coordination procedure:

      1. Resolving bottlenecks in project schedule.
      2. Continued interdisciplinary coordination.
      3. Design checks and approvals.

- Detailed specifications.

- Documentation.

- Monitoring engineering costs and progress.

- Staffing.

- Preparation of subcontract's technical packages.

The extent of the Project Manager's participation depends on the option followed to execute the detailed engineering. Under the single contractor approach, hands-on participation will be minimal, while in-house engineering will result in the most work for the Project Manager.

- Detailed engineering could not be performed without clear and detailed coordination procedures defining the scope of work and responsibilities assigned to each participant. When engineering is performed by a single engineering firm, large or small, the contractor will do what it always does and issue its own internal job coordination document. The participation of the Owner's Project Manager will be limited to reviewing the contractor's procedures and ascertaining that they are consistent with the scope of the contract and do not generate additional work.

  When the engineering is executed in-house or through several engineering firms, the burden of coordination falls squarely on the Project Manger's shoulders. The Project Manager must prepare and issue a coordination procedure and insure that all concerned comply with it. Failure to do so would have a negative effect on the project.

- Some owners only have general specifications and rely on engineering contractors to provide their own detailed project specifications. When the engineering is done in-house, they would use the specifications from some recent project in the same location. In that case, it is the Project Manager's responsibility to review and update them and, above all, make sure that they provide quality and safety consistent with the project requirements without unnecessary gold plating.

- Formal documentation of any communications affecting the execution of the work, especially the scope, is an absolute must to good project management. Normally, engineering contractors are very good at it, but it is up to the client's Project Manger to ascertain that the confirmation memos and minutes of meetings are accurate and truly reflect what was discussed and that proper emphasis is placed where it belongs.

When engineering is done in-house or by multiple firms, the Project Manager must take a proactive approach and assume the leading role in documenting relevant project activities, especially those related to defining changes in scope and/or schedule.

- The Project Manager must have hands-on participation in the monitoring of engineering costs and progress even if the contractor is using its own methods. In all cases, the Project Manager must establish control systems to make independent checks of the contractor's reports and evaluations. Chapter 10 includes procedures and forms that will enable the Project Manager to perform the control functions.

- Each contractor executing the detailed engineering will manage its forces and allocate them to the various active projects. The Owner's Project Manager must be alert to insure that the project is neither over-nor understaffed, nor shortchanged in the quality of personnel. It would be wise for the Project Manager to review and approve on a regular basis the personnel assignments with the contractor's Project Manager.

- When engineering is done in-house, the Project Manager must be continuously evaluating and making accurate forecasts of the remaining work to the appropriate local engineering managers so that they can make the necessary arrangements to meet the project staffing needs.

- The scope of work of any engineering firm doing the detailed engineering should include the assembly of a technical package to select construction contractors on a competitive basis. However, where the engineering is done in-house, it is up to the Project Manager to do that work.

CHAPTER 8
PROCUREMENT

8.1 Overview

On major projects, the selected EPC contractor handles the procurement function: purchasing, expediting, inspection, receiving and accounts payable. All the equipment and some of the critical bulk materials are purchased by the contractor. Bulk materials are generally supplied by the construction subcontractors as part of their scope of work.

Occasionally, as in the case of long delivery items, the Owner initiates the procurement process, (requisitioning, bidding and even placing of purchase orders) to insure timely completion of a project. Once a contractor is selected, the purchase order will be assigned to the contractor as part of the scope.

On small projects, procurement could either be assigned to the contractor doing the detailed engineering or performed by the Owner. In-house execution will undoubtedly minimize out-of-pocket expenses and seems like the desired option. However, the approach will only be viable for very small projects or on locations where an experienced and well-staffed purchasing department is available.

When procurement is performed in-house, the Project Manager assumes a very serious responsibility. Procurement involves much more than just placing purchase orders.

- Vendors must be screened.
- Approved vendors lists must be prepared.
- Certified vendors' drawings must be expedited for timely submittals.
- All vendors' activities must be expedited.
- Vendors' shops must be inspected and fabrication techniques approved.
- Equipment has to be inspected, tested and accepted.

- Equipment must be shipped and delivered.

Failure to perform these activities within the project time frame will result in schedule delays and a corresponding impact on engineering and construction costs. A compromise option, worth considering, would be to carry the in-house work up to the issue of purchase orders and assign all the follow-up and delivery responsibility to the engineering contractor.

If procurement is to be done in-house, a qualified purchasing agent must be assigned to the project team. It would be a grave mistake for the Project Manager to do purchasing personally. Effective procurement requires a tremendous amount of paper work and record keeping, including:

- Preparation and issue of inquiries.
- Bid analysis documentation.
- Communication with bidders/vendors.
- Procurement logs:
  1. Requisitions.
  2. Purchase orders.
  3. Inspection and acceptance.
  4. Invoices/payments.
- Claims processing.

In the absence of a contractor, all the paper work has to be done in-house and it would be impossible for the Project Manager to keep up with it without abandoning other duties and losing the overall perspective.

Although the guidelines in the following section were intended for project managers working with general contractors, they are also appropriate for small project managers.

8.2 Guideline for Purchasing

Inquiry

- Bidder list must be based on qualifications and ability to perform. Arbitrary restrictions will limit cost-saving opportunities.
- Specifications must be complete and sufficiently detailed to insure meaningful bids.
- Avoid overspecifying. It may unintentionally exclude less expensive but acceptable equipment.
- Search market to determine which vendors are hungry.

- Prequalify bidders of special equipment. This will insure acceptable bids.
- Premature inquiries increase home office costs and probably will not save any time.
- Allow reasonable time for bidding. Rush bids tend to be incomplete and will require more analysis and clarifications.
- Requests to extend bidding deadline may be an early indicator of future troubles. Vendor may be getting into more work than it can handle. It could also be an indication of unreliability.
- Competitive bidding is highly desirable, but under certain circumstances sole sourcing may be the right way to go.
- Sole sourcing must be justified and documented by the project manager or whoever requests it.
- Request firm price quotations, including unit prices for potential changes (nozzles and miscellaneous attachments) and spares, if required.
- All contacts with bidders/vendors should be channeled through the purchasing agent.

Technical Bid Evaluation

- The basic proposal must specifically respond to the inquiry and satisfy the technical specifications.
- If economically attractive alternatives are offered, determine whether they are technically acceptable.
- Evaluate operation and maintenance costs as well as potential impact on layouts, piping and spare parts inventories.

Commercial Bid Evaluation

- Make sure that costs for extras are properly segregated.
- Place a value on the operating and maintenance costs, as well as intangibles.
- If delivery is not acceptable, determine cost of improvement, if viable.
- If delivery is critical, ascertain feasibility of proposed schedule.
- Bids must be compared with budget estimates. The cause of substantial deviations must be determined.

- When the low bidder is not the recommended bidder, the reason must be justified and documented. A notation in the bid summary will suffice in most cases.

Purchase Order

- Make sure that all changes that have been negotiated are properly incorporated.
- On critical items, make sure that verbal confirmations are documented immediately by telex or fax.
- If the purchase order involves start-up and/or check-out assistance, make certain that the appropriate documents and information are included with the purchase order: insurance certificate, travel policy, site conditions, etc.
- Do not issue open-ended purchase orders.

Inspection and Expediting

- Define frequency and extent for every item at the beginning of the job.
- Beware of contractors with large inspection and expediting groups; unless controlled, they may overdo it.
- Force contractors to do their job. Don't do it for them. You should not relieve them of their responsibility.

8.3 Expediting and Inspection Criteria

Selecting a vendor and placing the purchase order is only part of procurement, usually the easiest part. If the article is not delivered on time or does not meet specifications, the entire procurement effort will be wasted and the project may not be completed within schedule and budget.

- Purchase orders must be expedited to insure timely delivery.
- Fabrication of equipment and materials must be inspected to insure quality.

On projects handled through a general contractor, procurement (including expediting and inspecting) is performed by the contractor under the supervision of the Owner's Project Manager. Frequently, the Owner's specialists will also participate in the inspection of critical items.

On projects handled under the small project approach, expediting and inspection are handled differently. Expediting is handled by whoever is doing the procurement, while the inspection is almost invariably handled by the Owner's specialists.

Large contractors usually have expediting and inspection departments and, naturally, they would like to keep them busy all of the time, especially on reimbursable contracts. If left to themselves, being human, they could be tempted to expedite and inspect every nut and bolt in the project.

Regardless of the project execution approach, the Project Manager must sit with the contractor and/or the in-house specialists, early in the project, and prepare a sensible program for selective expediting and inspection. Doing so will insure quality and schedule without unnecessary costs.

Expediting

Expediting of bulk materials and standard and/or off-the-shelf items can normally be done through periodic telephone follow-ups by the purchasing agent or even an inexperienced clerk. Expediting of large and/or critical equipment involving custom engineering and fabrication requires field expediting through visits to the vendors' office and fabrication shops by qualified personnel. Contractors normally use field inspectors or, sometimes, design engineers to do field expediting. The Owner would use, depending on the type of problem, either a CED specialist or the Contract Engineer. On a difficult expediting visit, the Project Manager is frequently called upon to lend weight to the undertaking.

The field/shop expeditor is expected to:

- Ascertain that the vendor prepares meaningful schedules and verify their viability.
- Concentrate initial efforts on early completion of vendors' drawings requiring approval.
- Obtain vendors' procurement schedules for long lead or critical items and verify their consistency with promised delivery dates.
- Obtain unpriced copies of vendor's suborders. For major subcontracted items, obtain permission from the vendor to expedite and inspect at sub-vendors' shops.
- Review shop fabrication schedules and observe progress of work in the shops and receipt of outside purchased equipment. Evaluate and determine if the shop scheduled versus actual work performance by the manufacturer is meeting the overall schedule.
- Continually evaluate the vendors' schedule and progress of work to discover potential problem areas.

- Apply intensive expediting effort on critical problems that may affect the delivery schedule.
- Prepare reports following each contact with the vendors to inform the Project Manager of progress being made and of problems requiring special attention.

Inspection

Inspection entails much more than examining a piece of equipment before shipment. It includes:

Pre-Purchase Inspection

- Survey vendor facilities to determine their qualifications. Capabilities assessment includes engineering, shop equipment, fabrication techniques, welding certification and QA/QC systems.

Post-Purchase Inspections

- Verification of materials of construction.
- Verification of all pre-weld and post-weld heat treatment to be used.
- In-process inspections, such as fit-up prior to welding, back gouging, in-progress dimensional checks and non-destructive examination (radiography, etc).
- Pressure and tightness tests.
- Performance tests on rotating and mechanical equipment.
- Final inspection for cleaning, preparation for painting and preparation for shipment.
- Review and acceptance of all required documentation.
- Release for shipment.

Inspection must be performed by qualifed personnel. Large contractors usually have a staff of qualified inspectors, while smaller contractors would use their mechanical or electrical engineers. Even when inspection is handled by a contractor, the Owner must retain the option to do an independent inspection or do it together with the contractor.

The Project Manager is not expected to perform any inspections but only to have sufficient knowledge of the inspection process to ascertain that it is properly done. Obviously, inspection has to follow a selective program. It would be a serious misuse of resources to subject every piece of equipment to a rigorous inspection procedure. As already mentioned, the Project Manager must develop a sensible program with the contractor and/or the CED specialists.

The following equipment would normally be included in an inspection and testing program:

- Tanks and pressure vessels.
- Heat exchangers.
- Processing equipment (dryers, mills, filters, mixers, centrifuges, etc.).
- Compressors and turbines.
- Special pumps.
- Custom or special valves.
- Boilers/furnaces.
- Packaged equipment (equipment consisting of several components, mounted in a frame and piped and wired together).

Performance Testing

Certain equipment will require testing either in the vendors' shops or in the field after installation. The degree of testing to be performed should be established prior to issuing the purchase order.

Testing may be simple no-load running to establish vibration levels and the functioning of auxiliary equipment, such as lubricating systems. It could also be an extended full performance test to be performed either at the vendor's shop or in the field. When full-scale performance tests are specified, detailed procedures should be prepared. Frequently, these procedures are available from organizations such as ASME, API and others.

CHAPTER 9
CONSTRUCTION MANAGEMENT

9.1 Overview

In a normal project, approximately 50% of the cost is incurred in the field:

- Construction:
  1. Bulk materials.
  2. Labor.
- Field Indirects:
  1. Labor related.
  2. Supervision and quality control.
  3. Construction equipment rental.
  4. Temporary facilities.
  5. Rework.
  6. Construction management.
  7. Other.

Additionally, the construction phase probably involves more risks, natural and legal, than any other project phase:

- Weather.
- Accidents and injuries.
- Physical interferences.
- Contractual disputes.
- Labor disputes:
  1. Plant personnel.
  2. Contractor personnel.
- Labor productivity.

Although some costs may be controlled by good project planning and engineering, and risks minimized by transferring them to contractors through intelligent contracting, there remain sufficiently undefined areas to make construction a phase of great opportunities and risks. It is up to the Construction Manager to identify cost saving opportunities as well as risks implementing the former while avoiding the latter. The quality of construction management can make or break a project.

Except in very small projects where construction is done with maintenance personnel, construction is always performed by outside contractors following any of the options discussed in Section 9.2, Construction Options.

The construction management function can be performed, depending on the construction option applied, by the construction contractor (G.C.), by an independent construction manager (C.M.) or by the Owner.

In all cases, the Owner's Project Manager has the ultimate responsibility of coordinating all parties involved and insuring that their work conforms to specifications and contractural conditions.

The purpose of this chapter is to:

-   Discuss the various options open to performing construction and construction management.
-   Give the Project Manager insight into the activities that take place as part of construction management.
-   Provide the Project Manager with guidelines and tools to do an effective job as C.M. on small projects.

9.2   Construction Options

Construction can be done using several approaches, most of which have been employed by the author on various projects:

*   As a continuation of a reimbursable EPC contract where the contractor is responsible for the construction using either:
    -   Direct hire.
    -   Subcontractors.
*   As a separate reimbursable construction contract where the contractor is responsible for the construction using either:
    -   Direct hire.
    -   Subcontractors.
*   As a separate, single, competitive lump sum contract.

- Through a construction management agreement where the Owner retains a C. M. to act as its agent to coordinate and manage several construction contracts awarded directly by the Owner.
- Through multiple competitive lump sum contracts managed directly by the owner who thus becomes in fact the C.M.

The first two options are more suitable for large projects. Lately many owners have been shunning away from the direct hire situation, and reimbursable EPC contractors are specifically instructed to perform the work through competitive lump sum subcontracts. Doing construction with a single, competitive lump sum contractor would be the ideal option for all projects if sufficient time were available to complete the design and prepare meaningful bids. Unfortunately, that never seems to be the case, especially in large and complex projects.

It is usually easier to arrange for a single, competitive lump sum contract on a simple, small project requiring less engineering and shorter bidding time. The only caveat to this option is that the chosen contractor must be able to perform most of the work without help from subcontractors. The use of subcontractors will add a layer of costs that need not be incurred if the Owner has the capability to do the construction management.

The construction management option of an independent C.M. does not seem to offer any advantages to either large projects or small ones. It is essentially the same as the first two options, executed through lump sum subcontracts, with the disadvantage that the Owner must retain a C.M. and still assume the responsibility instead of the contractor. On small enough projects, the Owner should act as its own C.M. and save construction management fees.

In many companies, it is normal that, once engineering is complete, the Project Manager will follow the job to the field to look over the contractor's shoulder with some construction supervisory assistance from either the plant or CED staff. Under the small project approach, the Project Manager and/or the Field Engineer act as C.M. and work directly with the construction contractors. When acting as C.M., the owner is assuming responsibility and risks normally assigned to the General Contractor. The extent of the risks involved and their probability must be evaluated for each case and weighed against the potential savings.

9.3   Construction Management Activities

The construction phase of all projects includes two well-defined sets of activities, actual construction and construction management.

Actual Construction

The actual construction comprises a variety of activities involving work by different specialized craftsmen:

Civil Work

- Site preparation.
- Underground mains and sewers.
- Foundations.
- Structural steel.
- Buildings.
- Painting.

Mechanical Work

- Relocations and modifications.
- Equipment setting.
- Piping.
- Instrument installation.

Electrical Work

- Power and lighting.
- Instrument wiring.

Specialty Work

- Insulation.
- Fire protection.

The work is performed by contractors usually specialized along the craft lines. Few contractors have the necessary breadth to perform all crafts directly; however, some have capabilities in one or two of the major crafts plus the expertise and resources to subcontract the rest. In doing so, they become General Contractors.

Construction Management

Construction Management involves the coordination, control and management of the contractors performing the various construction activities as well as providing them with a clean and safe environment. The person, or entity, performing this function is called the Construction Manager (C.M.).

A very important goal of construction managment is to promote a safe and pleasant environment conducive to a high level of performance by all contruction personnel.

As mentioned before, the C.M. function could be performed, depending on the construction approach, by the general contractor, an independently retained C.M. or by the Owner. In the latter option, the Project Manager becomes also the Construction Manager. Construction management entails many other activities, some of which are the responsibility of the contractors directly performing the work and others are the responsibility of the C.M.

Contractors

- Staffing.
- Payroll.
- Day-to-day scheduling.
- Direct craft supervision.
- Provide their temporary facilities.
- Craft training when required.
- Provide construction equipment and tools.

Construction Manager

Even when construction management is being performed by others (a G.C. or a C.M.), the Project Manager must have hands-on participation in some of these activities and supervise and spot check all of them. Those requiring special attention are marked with an asterisk.

- Layout temporary facilities:
  1. Parking.
  2. Offices.
  3. Shakedown areas.
  4. Storage areas.
- Develop site agreements with local unions.
- Enforce site safety and substance abuse policies.*

- Communicate potential operating hazards.*
- Organize and implement workers motivation programs.*
- Investigate and document accidents.*
- Provide safety indoctrination.
- Do activity checks.*
- Coordination:
   1. Among contractors.
   2. With plant.
- Prepare and monitor overall schedules.
- Perform progress measurement.*
- Receive, store and control Owner's purchased equipment and materials.
- Site security:*
   1. Gate control.
   2. Head counts.
- Quality control:
   1. Welders' qualification.
   2. Witness tests.
   3. Accept completed work.
- Overall housekeeping.
- Change order control:*
   1. Estimating.
   2. Maintain logs.
   3. Revise subcontracts.
- Provide and check coordinates, base elevations and locations.
- Verify layouts.
- Review invoices and approve payments.*
- Communications and documentation.
- Prepare emergency response plan:*
   1. Hold regular coordination meetings.
   2. Issue minutes of meetings.
   3. Keep daily logs of all relevant field activities.

9.4   The Project Manager as Construction Manager

Overview

On major projects, the construction management function is discharged either by the G.C. or by an independent C.M. specifically retained for the project. On small projects, the Owner is the Construction Manager. The C.M. activities are performed by the Project Manager and/or a Field Engineer.

All of the C.M. activities mentioned in the previous section are important and should be performed. However, those related to the administration of construction contracts, receiving equipment and field safety are of the utmost importance and must be performed and documented in order to protect the Owner's interests from potential contractual and/or labor disputes and claims stemming from the construction work.

When acting as C.M., the Project Manager or the Field Engineer must pay very special attention to the following areas:

- Holding weekly meetings with the contractor(s) and documenting all discussions and agreements through minutes.
- Documenting all discussions concerning contractor's performance.
- Enforcing safety rules and substance abuse policies.
- Investigating and documenting all work-related accidents and near misses.
- Documenting and following up on equipment receiving exceptions.
- Reviewing, approving and documenting extra work orders.
- Keeping daily logs of relevant field activities.

If the project is competed without any major mishaps, all the above may seem like an exercise in futility. However, should a contractor submit a claim for liquidated damages or a worker sue for a long-forgotten work injury, proper documentation would save the Owner a lot of aggravation and hundreds of thousands of dollars.

Initial C.M. Activities

The initial construction activities require personal attention from both the Project Manager and the C.M., even when the work is performed by contractors:

- Good relations must be established with the local unions.

- Rules have to be established to control the execution of the job and the coordination with other activities, plant operators and/or other concurrently active projects.
- A security system must be established.
- Temporary construction facilities must be set.

The contractor(s) doing the work has the direct responsibility of dealing with the local unions and/or labor pools. However, the Owner's Project Manager can insure that pre-job meetings are held with the local labor leaders and that reasonable site agreements are set to promote smooth relations throughout the project.

Project Procedures

The coordination of project activities with plant and/or other projects' activities is the responsibility of both the Project Manager and the C.M. They must promote and chair meetings with all interested parties to familiarize everyone with the existing plant rules and develop a set of project procedures consistent with them.

The following is checklist of the information that should be obtained prior to developing the project procedures:

- Who in the plant is the main contact?
- Plant safety procedures - specific hazards and dangers near the project and the field office.
- What is the procedure if a construction worker is injured on site?
- What are the security procedures for engineers and contractors?
- When are permits required? What are the types and how many are required?
- Listing of contacts for plant/corporate engineering project team members and contractors.
- Where in the plant are the contract personnel allowed to go for breaks or lunch?
- What are the area and plant working hours?
- How are fires and accidents reported?
- Which plant facilities, if any, would be available to the project?
- General plant dangers.
- Plant phone system.
- Local union holidays.

- Who is allowed to operate valves, switches, etc., in this plant?
- What contacts must be made when working late, weekends or holidays?
- What are the preferred plant procedures for handling final checkouts?

Security System

Establishing and implementing a project security system is a joint responsibility of all parties: contractor, Project Manager, C.M. and plant security.

Construction sites have frequently been a prime target for vandalism and theft. There are even cases of labor disorders ranging from incursion to assault. For these and other reasons (e.g., public safety), the security of a construction site is a very important consideration. While usually not requiring Fort Knox-type protection, the Rittenhouse Square-type (wide open and unlimited access) is totally unacceptable.

Some of the considerations when designing a site security system are as follows:

- Is local law enforcement adequate to handle any anticipated problems?
- Is local law enforcement sympathetic with owners, unions, etc?
- Can a designated gate be established for construction use?
- Should trailers, offices, storage areas or staging areas be lit at night?
- Fuel tanks should be locked and gas bottles chained.
- Material storage trailers should be locked.
- High value materials should be stored in secured locations.
- Is 24-hour guard service needed?
- Limit afterhour site access.
- All vehicles leaving the site should be subject to search.
- Limit the number of vehicles allowed on the plant site proper.
- Use gate passes for all materials leaving the site.
- Use high security locks with non-duplicable keys.
- Union and non-union parking areas should be separate.
- Each employee should have some means of identification, i.e., badge, colored hat, etc.
- All emergency phone numbers should be posted at each phone.
- Are alarm devices on trailers or offices worthwhile?

- Do the security measures taken focus on deterring or delaying criminal actions?
- Is present fencing adequate?
- Are temporary lights required?
- Are firefighting equipment and services available quickly?
- What is the history on frequency and degree of labor militancy?
- If a fence is part of the project, does it make sense to erect it very early?
- Is there adequate lighting at the work site?

A rational layout of the temporary facilities (offices, change houses, shops, parking lots, layout areas, etc.) enhances labor productivity and has a favorable impact on project costs and schedule. One of the main responsibilities of the C.M. is to lay out the temporary facilities in an efficient manner to minimize the distance the workers have to walk from the gate to the work areas and design traffic patterns to avoid jams and bottlenecks.

Recommended Field Reports/Logs

One of the responsibilities of the Construction Manager is to generate, and/or insure that the subcontractors do, sufficient documentation to:

- Monitor and control field activities.
- Keep management informed.
- Collect meaningful data for future projects.

The following list of reports and logs is probably the minimum required to meet the above objectives:

- **Daily Force Report.** Each subcontractor must submit a daily force account broken down by crafts. The Construction Manager must keep a running total for each subcontractor for inclusion in the final project report and for future reference.
- **Look Ahead Reports.** Each subcontractor must submit every week a list of activities planned for the next two weeks with projected staffing. The Construction Manager must compare this information with the overall schedules to gauge subcontractors' performance and promote corrective action if necessary.

- **Progress Reports.** All subcontractors must submit at least every two weeks progress reports comparing actual versus scheduled progress. The Construction Manager should have the means of making quick, independent checks.

- **Progress Meeting Minutes.** The Construction Manager must issue the minutes of the regular construction coordination meetings.

- **Field Logs.** The Construction Manager must keep a field log documenting all important field events as well as verbal communication and decisions concerning execution of the work and relations with the subcontractors.

- **Change Order Log.** The Construction Manager must maintain an up-to-date log of all field changes, approved as well as pending.

- **Accident Reports.** Each subcontractor must submit reports on every lost time accident. The Construction Manager must participate in the investigation of major accidents and insure that reports on all accidents are promptly filed and kept as protection against potential future legal action.

- **Acceptance Reports.** The Construction Manager must document and file all final equipment inspection and acceptance.

- **Construction Status Reports.** The Construction Manager must issue every month, for management information, a complete report on all construction activities with cost and scheduled evaluation for compliance to the AFE objectives.

9.5   Influence of CICE on Construction Management

In the late 1970's, the Business Roundtable sponsored an extensive study to determine the cause of the diminishing productivity being experienced by the construction industry in the United States and to propose solutions to stop and reverse the trend. A team was formed with representatives of prestigious universities, contractors and owners from several sectors of industry. As a result of this concerted effort, the Construction Industry Cost Effectiveness report, now widely known as CICE, was issued.

The CICE report identifies and proposes remedies for a series of specific problems directly related to the decline in productivity. Those directly related to the field construction activities are considered to have the most important impact on project costs.

- Poor safety practices.
- Poor construction management practices.
- Lack of worker motivation.

These are areas where all project managers, whether acting as such or as construction managers, must assume the leadership and exert their power to influence the outcome of the field activities and insure successful completion of the project.

The Construction Industry Institute (CII) was established for the purpose of implementing the CICE recommendations and is continuously following up and expanding upon all CICE-related issues. Every Project Manager, active and/or fledgling, must make a point to study and become fully familiar with its recommendations.

9.6   Co-Employmentship

The standard contract usually indicates that the Project Manager speaks and acts for the Owner. When the Owner's representative gives working instructions directly to a craftsman, a situation of co-employership is set up. The workers get their check from one party (the contractor) but their working instructions and directions from another (the Owner). Thus they, in effect, have two employers. The courts hold both employers responsible in case of accident or injury. In cases where co-employership is claimed, both the Owner and the Project Manager are open for legal action.

The best way to avoid this situation is to give all directions through the contractor. It is best for the Project Manager, and other Owner's personnel, not to give directions directly to craftsmen. Even an instruction that appears to be the most harmless and innocent can lead to a legal case of co-employership.

The only time it is proper to give direction to a worker, bypassing the chain of command, is when you see that worker in eminent danger of health or life and immediate action is required. Other than that, always work through the proper chain of command.

CHAPTER 10
PROJECT CONTROL

10.1 Thoughts on Project Control

Project control is the only activity that spans all other projects execution phases.

- Project control is practiced during scoping, process design and execution planning.
- Project control is exercised when establishing contracting strategy and selecting contractors.
- Project control is also practiced during all phases of engineering, procurement and construction.
- Project control is the driving force behind progress monitoring and contract administration.
- Project control entails quality, cost and schedule controls.

PROJECT CONTROL IS INTRINSIC TO GOOD PROJECT MANAGEMENT AND IS ACHIEVED THROUGH THE EFFECTIVE EXECUTION OF EVERY PROJECT ACTIVITY.

NOTE: Quality, defined as conformance to established requirements must be monitored by process and specialist engineers with minimal direct participation from the Project Manager. Consequently this chapter addresses only cost and schedule control.

Many people confuse cost tracking and/or accounting with cost control; that thinking is far from the truth. The greatest opportunities for real cost control and substantial cost reductions are found in the early project definition and planning stages - site selection, process design and execution planning. Cost tracking and accounting come into play after the project has been defined and the major costs set. Cost tracking is an important activity that must be performed in order to detect variations, take corrective action, if possible, and make accurate forecasts. However, the opportunities for cost reduction are very limited.

Cost accounting is of little use to the Project Manager other than gathering cost information for future projects. Even that may not be true since on many occasions the accountant's methods are not consistent with the project estimating and control needs.

Costs are set very early in the project and the areas that offer greatest opportunities for real cost savings are project definition and execution planning.

- The decisions made during the initial stage will fix the minimum cost of the project - design criteria, execution strategy, contractor selection, plant layout, project specifications, etc.
- Good planning and scheduling will minimize the project duration and enhance productivity both during engineering and construction.

An alert Project Manager can realize real cost savings in these areas by being inquisitive, having a healthy skeptical attitude and a good sense of cost and value. The Project Manager does not have to be an expert in any particular field but must have a basic understanding of all disciplines involved in project execution.

Project Managers are charged with the responsibility of insuring that projects are executed within the scope, budget and schedule established in the AFE. Project control is then their prime responsibility.

- Project control is a continuous and live activity requiring constant attention. It cannot be performed effectively on a spot or part-time basis.
- In small and medium sized projects that will not support a full-time Cost Engineer, the Project Manager must take an active hands-on role in cost control.
- In doing so, the Project Manager must seek the advice of the cost control specialists to insure their participation in setting the project control systems and checking certain key activities.
- Large projects can support and do require a full-time Cost Engineer to work very closely with and keep the Project Manager continuously informed of all variations and/or potential dangers.

- Good project control implies both thought and action. It means:

- **Anticipating**	(Thought)	|
- **Avoiding**	(Action)	|
- **Recognizing**	(Thought)	| **TROUBLE, ON TIME !!**
- **Correcting**	(Action)	|

If this is done, the project will be under control. If it is not, the project will run into trouble no matter how elaborate the control procedures are or how thick and detailed the cost report is.

10.2 Project Control and the Project Manager

As previously mentioned, project control is a continuing and live activity that requires constant attention from the Project Manager. It begins at inception and must continue throughout the life of the project until final completion.

THE PROJECT MANAGER EXERCISES CONTROL BY TAKING A DELIBERATE APPROACH TO INSURE THE EFFECTIVE EXECUTION OF ALL PROJECT ACTIVITIES.

Cost Control

The greatest cost saving opportunities are found during project development, definition and execution planning. They include:

- Site selection.
- Phase 0/Phase I.
- Project Execution Plan.
- Contracting.

These activities are normally performed in-house, thus allowing those involved the greatest opportunities of influencing costs, either favorably or adversely. It behooves the Owner's management to implement a protocol of checks and balances to guarantee:

- The application of uniform standards and procedures.
- The participation of people with solid project management/engineering experience and good cost background.

- Reviews and approvals at proper levels.

Ideally, the assigned Project Manager should be the one personally providing project execution and cost input during the initial phases. If that is not possible, somebody else with project management/engineering experience should be assigned until the Project Manager is available.

Once the project is defined, cost control must continue, mainly as a tracking and monitoring activity during the execution phases of:

- Engineering.
- Procurement.
- Construction.

Although the cost saving opportunities during project execution are limited, the tracking and monitoring are very important activities. Through them the Project Manager can:

- Detect variations early enough to . . .
- Take corrective action if feasible, or in any case, . . .
- Make accurate forecasts.

In a conventional project, the execution stages are performed by an engineering and/or construction contractor who acts as general contractor (G.C.) and has its own so-called cost control system. They are actually tracking and/or monitoring procedures. When the execution is contracted on a cost reimbursable basis, the Owner is interested in both cost tracking and progress monitoring. In lump sum contracts, the Owner's concern should focus primarily on progress monitoring.

Some of the contractors' systems are very sophisticated and all are good, if properly implemented; unfortunately, those systems are frequently only partially implemented and, occasionally, not implemented at all. It is the responsibility of the Project Manager to insist that contractors implement their systems.

It is also the responsibility of the Project Manager in cost reimbursable projects to insure that the sophistication and cost of the control system employed by the contractor is commensurate with the project requirements.

WORD OF CAUTION: The best practice is to let the contractors use their own systems. Imposing an Owner's systems could be a grave mistake.

The cost control activities of the Project Manager during the execution stages are practically the same for both conventional and small projects. In any case, in order to exercise control of his project, the Project Manager must:

- Become the project team's cost conscience and act as devil's advocate in all matters affecting cost.
- Have a good sense of cost and be capable of making on-the-spot cost evaluations to minimize wasteful and time consuming studies.
- Challenge, and make the originator justify, any changes that imply deviations from the basis of the appropriation estimate.
- Make independent spot checks of the cost information supplied by contractors, subcontractors and other team members.

Schedule Control

As in project cost, the greatest opportunities to influence project schedule are found during the initial project stages. Cost and schedule are intimately related and the schedule corresponding to the minimum execution cost is the optimum one. Any deviations could have a negative cost impact.

Schedules are very frequently dictated by business considerations and, as discussed in Chapter 4, rarely coincide with the lowest execution cost. In extreme cases the dictated schedule may only be achieved at an outrageous cost. It is the responsibility of the Project Manager, during the planning stage, to:

- Warn management whenever the desired schedule is not consistent with the minimum execution cost.
- Develop cost effective schedule improving schemes.
- Prepare a realistic master project schedule and firm execution plan to support the appropriation estimate (Chapter 4 and 5).
- Work closely with the selected engineering contractor to ascertain that the final execution plan and project schedule are consistent with the goals stated in the AFE.

Once engineering is complete, the project moves to the field, and construction is contracted, preferably on lump sum basis. Schedule then becomes the prime responsibility of the construction contractor(s). However, the Project Manager must make independent progress evaluations to confirm that the work is proceeding according to schedule and no expediting action is required.

10.3 Control in the Early Stages

At the risk of being repetitive, it must be said once more that the greatest opportunities of real cost savings are found during the initial project stages.

It is in this period when experienced project management is most needed. Unfortunately, it is also when decisions are sometimes made without the benefit of

project management input. The probability of success of every project is greatly
enhanced when project management is invited to participate in the very early
stages.

Project managers should always be on the alert for instances where projects are
being developed without project management input and bring this to CED
management's attention.

The best contribution the Project Manager can make in the scoping and process
design stages is to:

- Come up with quick conceptual estimates to evaluate the cost impact of
  proposed alternatives.
- Evaluate possible construction methods.
- Generate and maintain a cost awareness in the design team.

The succeeding paragraphs include guidelines and tips applicable to the various
initial project stages.

Site Selection

Site selection is probably the activity with the single greatest potential cost impact.
The site factors affecting cost can go well beyond those affecting capital cost;
some, like those listed below, would also impact production and marketing costs.

- Raw materials cost.
- Utility costs.
- Accessibility to major markets.
- Local availability of qualified personnel.
- Living conditions for "imported" personnel:
  1. Housing.
  2. Schools.
  3. Hospitals.
  4. Entertainment.

Some of these factors fall outside the Project Manager's scope of work. They are
evaluated by the business groups and the decisions are taken by management.

When it comes to factors affecting the capital cost, it is the responsibility of the
Project Manager to develop the information and estimate, conceptually, the cost
impact of site factors such as:

- Local construction labor:

　　1. Availability.
　　2. Productivity.
　　3. Rates.
-　Availability and capabilities of local contractors.
-　Availability and proximity of utilities.
-　Site conditions:
　　1. Soil bearing.
　　2. Area drainage.
　　3. Topography.
-　Environmental requirements.
-　Permitting.

Phase 0/Phase I

During the Phase 0/Phase I design, the Project Manager can start controlling the ultimate project cost and avoid unpleasant surprises by:

- Providing on-the-spot assessment of the cost and schedule impact of the various process alternatives considered during process design.
- Tracking costs and making projections as the design progresses, thus providing sound criteria for timely action by the technical manager and, when required, Owner's management.
- Insuring that all factors with potential cost impact are taken into account so that they can be included in the cost estimates and considered in the project schedule.
- Helping the Technical Manager in the development of cost-effective layouts and equipment arrangements.
- Insuring that the specifications do not include unnecessary gold plating:

  - Excessive corrosion allowances and vessel thickness.
  - Ultraconservative materials of construction.
  - Superfluous block and bypass for control valves.
  - Unnecessary installed spare pumps when process allows for short unit shutdowns.
  - Structural design for concurrent adverse conditions.
  - Disproportionate instrumentation because the control system has the extra capacity.
  - Requests for needless contractor documentation.

Project Execution Plan/MPS

A well-thought out and timely issued project execution plan chooses the most cost effective route and avoids costly vacillations, false starts and unexpected changes in direction. It anticipates and avoids pitfalls and plans alternate routes for the blatant ones.

> THE RESPONSIBILITY OF PLANNING THE EXECUTION OF ALL PROJECTS, LARGE OR SMALL, FALLS SQUARELY ON THE SHOULDERS OF THE PROJECT MANAGER. IT IS A RESPONSIBILITY THAT SHOULD NEVER BE DELEGATED.

The minimum capital cost execution plan is usually one based on executing the entire project through a competitive lump sum contract based on a firm and accurate scope.

- This goal is rarely a practical one. In chemical plants, especially when the process is based on Owner's technology, the Owner wants to keep control of the extent and depth of the engineering performed to assure quality. The Owner also wants to maintain flexibility to explore design alternatives without having to hassle with the contractor over who should pay for any added costs. To meet these requirements, engineering should be performed on a cost reimbursable basis.

- Even if engineering is performed on a cost reimbursable basis, construction should always be done on a lump sum competitive basis through a single G.C. contract or through several single subcontracts managed either by the owner or by the engineering contractor acting as Construction Manager.

The best contribution that the project manager can make to project control in this stage is to develop a viable execution plan based on maximal use of competitive lump sum contracts, a realistic schedule and an accurate cost estimate.

In a nutshell:

> THE PROJECT MANAGER MUST BUILD THE PROJECT ON PAPER IN ORDER TO ANTICIPATE, AVOID, RECOGNIZE AND CORRECT TROUBLE ON TIME.

Contracting

Intelligent contracting reduces project costs and enhances the probability of success. The Project Manager must work closely with the Contracts Engineer to develop a contracting strategy that results in the selection of the contractor(s) most likely to do a quality job at a minimum cost.

Contracting construction on a lump sum competitive basis is a simple proposition even if engineering is not 100% completed. Once the bids have been qualified and found acceptable, the low bidder will provide the minimum cost.

Contracting engineering services on a cost reimbursable basis, for projects of any size, is a far more complicated undertaking, requiring subjective evaluation of many intangibles. Different contractors, depending on their size, organization structure and degree of sophistication, will require different hours to perform the same work. It is up to the Project Manager to sort out these variables and select the contractor most likely to execute a particular project at minimum cost.

- Large contractors normally have long chains of command and highly structured organizations. They will employ 10-20% more people than small contractors with flat organizations.
- Some small contractors also have long chains of command. Their hours will also be high.
- Contractors, large or small, committed to sophisticated project management and/or control systems may be incapable and/or unwilling to use simple methods that could be more cost effective and quite adequate for small projects. Their hours will be high.
- Small contractors may not be able to provide the staff required to do the work within the required time frame. Choosing them could be disastrous.
- If a large, inefficient contractor has a unique expertise required for a particular project, it will be worth trying to retain it as a consultant on the particular area and award the bulk of the work to a small, more efficient contractor.

The Project Manager can minimize engineering costs by doing the necessary homework to assist the Contracts Engineer in the bidder selection process. The Project Manager should:

- Prepare an estimate of home office hours.
- Determine the staff loading required at the peak of the engineering activity. A single project should not use more than one-third of the contractor's staff.
- Determine if a particular expertise is required for the project at hand.

Table 10.1 Engineering Execution Alternatives

Project Approach	Contractor Size	Work-hours per Equipment Item			
			Owner		
		Contractor	Eng	Purch	Total
Large	Large	690	0	0	690
Large	Small	540	0	0	540
Small	Large	491	0	35	526
Small	Single Small	364	0	32	396
Small	Several Small	306	0	30	336
Small	Maximum In-House	55 [1]	174	28	252

[1] Only Civil and Structural Engineering.

Table 10.1 illustrates the potential cost impact of the various alternatives for executing detailed engineering. It was developed with the engineering hours estimating procedure included in Chapter 13 and assumes that in the case of the small projects, the Project Engineer/Manager assigned to the job takes an active role in managing the detailed design activities as well as implementing project controls

10.4   Control in the Engineering Office

General

The basic design has been completed and an engineering contractor has been selected. The project is now ready to move into the contractor's office for detailed engineering and, in most cases, procurement. Normally the work will be done under any of the various options discussed in Chapter 6. Although the material presented in this chapter is specifically addressed to a cost reimbursable situation, the basic concepts are also applicable to lump sum contracts.

   The contractor is legally bound to perform the work according to the contract. It has also assumed the professional responsibility and will be liable for failures stemming from the mechanical design and/or execution provided under the contract. It has a contractual right to take the necessary steps and incur reasonable expenses to protect its liability.

   On the other side of the coin, the Owner's Project Manager is generally responsible for making the contractor fulfill the contractual obligations and insuring that the costs incurred are reasonable. However, in doing so, care must be taken to not abrogate the contractor's responsibility placing it on the Owner's shoulders. Such a step could have very grave consequences and should not be taken without consulting with management. Specifically, the Project Manager must:

- Ascertain that the contractor, as well as all members of the Owner's project team, fully understand the scope of the work and the cost and schedule objectives.
- Review and approve engineering budget hours.
- Approve all commitments, expenditures and changes.
- Insist that the contractor actually implement the design and control system promised in the proposal.
- Take active participation, through the contractor's Project Manager, in the planning of the work to insure that the engineering work will support lump sum competitive construction work.
- Make independent progress and productivity assessments.
- Take an active role in checking schedules and cost estimates.
- Instill a sense of cost consciousness in the project team.
- Discourage changes but, if they must occur, make sure they are approved at the right level.
- Direct the Owner's project team, including the part-time specialists, and insure their timely participation to provide adequate technical control and keep the job flowing smoothly and efficiently.

Neither engineering contractors nor the team specialists are omniscient, and the Project Manager should never hesitate to challenge and make them justify their position and/or recommendation. "We always do it this way" is not an acceptable answer.

Plant Layout

Plant layout is one area where real cost control can be exercised because of the impact on material and construction costs as well as on operating efficiency. The final layout, usually developed during the basic engineering stage, requires a high degree of experience and the participation of the entire project team as well as the contractor's specialists and Owner's operations and maintenance personnel.

Establishing a fixed layout early on in the project should be one of the Project Manager's prime objectives since this is a means of insuring effective and economical detailed engineering and early project completion. Achievement of these goals requires:

- Timely participation of all involved.
- Awareness of the costs involved in the various options.
- Adamant resistance to changes of any kind to the approved layout.

Many people believe that the layout of a chemical plant is an art. That may be so, but it is also the result of common sense and an understanding of the costs involved in executing the various alternatives that are usually available.

The cost conscious Project Manager with a good feel for costs will usually find cost saving opportunities in the development of plant layouts.

Detailed Engineering

Some cost saving opportunities may be found during detailed engineering, especially during the early stages when the project design procedures and specifications are being set. However, the contractor working on a cost reimbursable basis may not take the initiative to propose cost saving schemes. Therefore, the Owner's Project Manager must watch for those opportunities and have the contractor take advantage of them. This, of course, must be done without relieving the contractor from the responsibility of doing a proper job and keeping schedules.

- The choice of design presentation will have an impact on costs, especially if construction is to be subcontracted on a lump sum basis. If that is the case, engineering work must also be scheduled accordingly.
- Planning the engineering work so that mass purchasing can be done will not only result in fewer procurement hours, but also will be a factor in securing better prices.
- Overly conservative contractors' specifications can be very expensive. The Project Manager must challenge them aggressively.
- Asking the contractor to comment on the cost impact of special Owner's requirements may result in cost savings. In some cases the originator of the request may not be aware of the cost and may very well waive or relax the requirements, If the cost differential is still excessive, the requester should justify the expense.
- Timely participation of experienced construction personnel during the layout work could avoid future installation and operation problems and result in substantial savings during construction.

Here are some ideas that could be considered to reduce engineering and/or construction costs:

- Standardize the configuration of pile caps and foundations.
- Adapt design of structural concrete to shapes that minimize construction work - prefabricated elements, standard reusable forms, etc.
- Neat cut, excavate and pour foundations directly on the soil. Forming is expensive, concrete relatively cheap.
- Base structural steel design on standard modules.
- Consider using flexible steel tubing for small bore pipes and or shop bent pipe for large sizes.
- Standardize pump casings and piping around pumps.
- Send pipe designers to the field during construction to detail the routing of small-bore pipe, steam tracing, safety stations and service stations.
- Include as much detailed engineering as possible in the scope of competitive lump sum subcontracts and purchase orders for mechanical packages:
  1. Contract buildings on design-build basis.
  2. Have mechanical subcontractors do their own piping isometrics whenever feasible.
  3. Have vendors include as much piping and supports as possible on skid-mounted packages.

The detailed design involved will thus be done on a competitive lump sum basis and some savings will be realized.

Purchasing and Subcontracting

Effective purchasing and subcontracting can provide opportunities for realizing substantial cost savings. Savings can be realized through:

- Better prices resulting from intelligent competitive bidding and/or negotiating and subcontract administration.
- Better quality resulting from proper screening and inspection of vendors and subcontractors.
- Improved completion dates resulting from effective expediting.

Contracting construction work through lump sum competitive bidding can have a very significant impact in construction costs. However, in fast track projects, this requires a concerted planning and coordinating effort by the engineering procurement and construction groups to make it happen. The Project Manager must take a very adamant attitude in insisting that all groups display the utmost ingenuity and flexibility to insure competitive lump sum contracting without impairing the project schedule.

Purchasing practices and procedures are an integral part of the contractor overall organization and will not be changed by a client's Project Manager. The best, and probably only, manner to insure effective purchasing and contracting is to select a contractor with a good and lean organization and procedures.

If the selected contractor does not have a good purchasing department, about the only thing the Project Manager can do is to get help to reinforce supervision, and support the contractor organization if required. This is not the ideal way to run a job but may be the only way to minimize costs and protect the Owner's interest.

10.5 Control During Construction

The most effective way to minimize construction cost is to:

EXECUTE CONSTRUCTION ON A HARD MONEY COMPETITIVE BASIS, EITHER THROUGH LUMP SUM UNIT PRICES CONTRACTS OR A COMBINATION OF THE TWO.

Some contractors, especially large ones, will argue that this approach is not compatible with fasttrack schedules. This is far from the truth; a flexible engineering contractor committed to the competitive lump sum approach will always find ways to create competitive situations without hindering the schedule, especially if the Owner's Project Manager is adamant on the subject.

- Break down the work into discrete portions and plan engineering accordingly.
- Go through a bidding round when engineering is only about 60% done, continue engineering during the bid preparation and go through a second bidding round, with the two low bidders, when engineering is more complete and maybe a final negotiation with the low bidder and up-to-the-minute engineering.
- Have good definition of the "undefined" as well as the defined work.
- Ask for competitive unit prices based on preliminary take-offs and convert to lump sum based on actual quantities.
- Pre-order 80% of bulk materials normally supplied by the contractors in order to minimize mobilization time.

When construction is awarded on a lump sum basis, most of the risks as well as the cost control function is assumed by the subcontractors. The Owner's main cost control responsibility is limited to controlling extras and monitoring quality and progress. On small projects, the Owner's Project Manager would normally act as Construction Manager and be directly involved in the cost control activities.

Even on small projects, the Owner could delegate construction management to either the engineering contractor or to a Construction Management (C.M.) firm. Although this approach is more expensive, it could be dictated by the complexity of the field work and/or the lack of the Owner's resources at a given moment. Although the C.M. will bear the brunt of the control work, the Owner's Project Manager must sill conduct spot progress checks and be on the alert to avoid overstaffing of the C.M. team.

Whether construction management is being handled by a C.M. or in-house, the Project Manager must assume the leadership in enforcing the highest degree of safety standards, sponsoring workers motivation programs and, in general, promoting the implementation of the CICE and CII recommendations.

10.6 Control During Project Control

A common trap during project execution, especially when dealing with large EPC contractors, is control overkill. Many contractors will flaunt their "unique" control systems guaranteed to keep any project out of trouble. Most of these systems far exceed the requirements of small, even conventional, projects, and none will keep them out of trouble in the absence of qualified and experienced project management.

Sophisticated control systems are invaluable, and should be obligatory, in mega projects as well as in extensive turn-around work, where all activities must be performed with clockwork precision. However, for small projects, as well as many major ones, they increase execution costs without real benefits. It is up to the Project Manager to ascertain that the controls applied by the contractors are commensurate to the size and complexity of the project.

The Owner's Project Manager and Cost Engineer must have an active participation in setting up the Contractor's control system for cost reimbursable projects. This must be done at the very beginning of the contract work.

The cost and complexity of the system must be commensurate with the project requirements:

- If a simple manual system is adequate for a small project, there is no need for a sophisticated system.
- If the volume of information can be handled by the Project Manager, there is no need to assign a Cost Engineer.
- Unnecessary information will add to project costs and create confusion.

However, sufficient information must be provided to allow the Owner's Project Manager and/or Cost Engineer to spot check contractors' claims and make independent assessments.

10.7 Anatomy of a Project Control System

As previously mentioned, the greatest opportunities for real cost control and substantial cost reductions are found in the early project definition and planning stages. However, once the basic decisions have been made and their cost impact estimated, a control system must be established.

"Cost control system" is a commonly used misnomer since the decisions having the greatest impact on cost have already been made. "Project tracking" or "Project monitoring system" would be more appropriate. However, a good system, in the hands of an experienced Project Manager will provide more than tracking or monitoring; it is a tool for anticipating, avoiding and recognizing trouble so that it can be corrected on time. It is really a "project control system".

The prime purpose of a project control system is to provide an accurate, quantitative means of measuring:

- Performance - actual versus estimated costs, hours and productivity.
- Actual progress against the established targets for both cost and schedule (adjusted for major scope changes).

in order to:

- Detect problem areas in time to take corrective action.
- Identify positive influences and capitalize on them.
- Make valid and reliable schedule and cost forecasts.

The Owner's cost control system need not be as sophisticated as the contractor's. It is intended mainly to check the accuracy of the contractor's progress reports and cost and schedule projections or to control small in-house projects. A simple manual system judiciously used should be sufficient to meet the Owner's requirements.

An effective system, simple or sophisticated, must include procedures and provide tools to perform three basic functions:

- Cost tracking.
- Progress monitoring.
- Continuous cost and schedule evaluations and forecasts.

The prompt and accurate discharge of those functions provides project management with the necessary information to take timely corrective action to control cost and schedule and inform upper management of unavoidable deviations.

- The bases of accurate **cost tracking** are:

    - A cost estimate broken down in quantifiable discrete components that can be easily comprehended and evaluated by simple observation without going into the details of counting and tracking "nuts and bolts" (Section 5.4).
    - The discipline to track the cost components - quantities and unit costs and/or man-hours - at the completion of discrete portions of the detailed design of the various physical areas, process sections and/or disciplines.

- A **progress monitoring system** must include the following elements:

    - A valid and reasonably accurate list of activities to be performed, consistent with the cost estimate breakdown and "loaded" with the estimated manpower and/or cost for each activity.
    - A flexible procedure to assign consistent, relative values to the diverse activities based on any combination of the following parameters:
        1. Dollar value.
        2. Importance to project.
        3. Exposure (potential downside).
        4. Construction and engineering hours.
        5. Physical units - yd^3, ft, ton, etc.
    - An evaluation method, based on specific milestones, to objectively gauge the progress of each individual activity.

- Accurate **cost and schedule evaluations and forecasts** require:

    - Means of measuring the resources (dollars/hours) applied to each activity and the discipline of making continuous comparisons with the estimate.
    - Master schedules and short-term, itemized schedules and the discipline of tracking performance and making the contractors live up to their projections, especially the short-term ones. If a contractor cannot meet the goals of a short-term schedule, it will never meet the long range schedules. The Project Manager has to take immediate action!

- Bi-weekly performance (productivity analysis and cost and schedule forecasts).
- Long range and short-term staffing plans.
- Periodic reports.

The best cost control system per se is totally useless if not implemented. Cost control is a dynamic activity that requires participation and commitment from the entire project team, contractor's as well as Owner's. Control is exercised through:

- Continuous comparison, by the design leaders, between actual design quantities and control estimate quantities.
- Deliberate efforts to correct unfavorable deviations without impairing quality.
- Proper and adequate project team communications.
- Regular interdisciplinary coordination meetings.
- Comparison between vendors' bids and control estimate.
- Constant comparison of progress with the schedule.
- Continuous monitoring and expediting of vendors' progress.
- Continuous monitoring of subcontractors' performance.
- Regular cost tracking meetings.
- Regular coordination meetings with subcontractors.
- Timely reporting.

10.8 In-House Cost Tracking

As mentioned in the previous section, the only requirements for effective in-house cost tracking are:

- A cost estimate broken down in quantifiable discrete components.
- The willingness and personal discipline to track and compare the original preliminary take-offs and costs with the latest ones as the design develops and firm commitments are made through purchase orders and subcontracts.

The semi-detailed estimating procedure presented in Chapter 13 provides sufficient breakdown for adequate cost tracking by the Owner's Project Manager. The willingness and personal discipline must be provided by the Project Manager. The following cost components, all developed with the aid of the semi-detailed estimating procedure, can be easily tracked by the Project Manager.

- Equipment count - from equipment list.
- Equipment cost - from purchase orders.
- Motor count - from equipment list.
- Plot size - from plot plans and arrangement drawings.
- TDC Control Points - from P&ID's.
- Field Instruments (balloons) - from P&ID's.
- Concrete take-offs - from design drawings.
- Structural steel take offs - from design drawings.
- Building size - from plot plans.
- Heat tracing required - from P&ID's.
- Insulation required - from P&ID's.
- Piping take-offs - from bills of material and line lists.
- Piping materials costs - from purchase orders.
- Engineering hours - from EPC contractor forecasts.
- Field indirects - from EPC contractor estimates.

The cost tracking activity could become cumbersome and expensive when construction is executed using direct-hire rather than lump sum contracts. In that case, labor productivity becomes an important cost factor and must be monitored closely to determine trends, try to make corrections and make accurate forecasts.

10.9 In-House Construction Progress Monitoring System

General

All lump sum contractors should be contractually obligated to impose and implement a progress monitoring system. A good system need not be expensive and should not increase contract costs, no matter what a contractor may say.

Not withstanding whether the contractor implements its own system, it behooves owners to develop in-house systems for confirming contractors' claims quickly and with reasonable accuracy. The preparation and implementation of a progress monitoring system require the following elements:

- An activity breakdown in discrete components that can be easily comprehended and evaluated by simple observation.

- A flexible procedure to assign consistent relative values, expressed in physical units, to the various activities that reflects, in a rational manner, not only work hours but also their real value and/or importance to the project.
- An evaluation method based on specific and clear milestones to objectively gauge the progress of each activity.
- A firm project execution plan and MPS as outlined in Chapter 4.

Activity Breakdown

A semi-detailed cost estimate prepared with the procedures and guidelines included in Chapter 13 could easily be dissected into a multitude of comprehensible construction activities that could be integrated, at the Project Manager's option, into a manageable number of discretely sized units. Every single activity could be assigned consistent relative values based on field hours

- Equipment erection - by area or by item.
- Foundations - by area or by item.
- Steel structures - by area or by item.
- Instrumentation - by area or by item.
- Piping fabrication - by area or by line.
- Piping erection - by area, by system or by line.

The proposed estimating system has the capability of breaking down the work activities to the level of each instrument and each pipeline. This level of detail is not required in an in-house system and would defeat its main reason for existence, confirming contractors reports quickly and accurately. A system based on 100 to 200 activities, depending on project size, would provide a good level of accuracy and, once set up, could be easily implemented. The maximum value of each activity should be kept below 5% of the total.

Value System

Once the work has been broken down in activities, relative values must be assigned to each of them in a uniform manner that can be related to simple physical units. The values could be assigned either as work units (W.U.'s) related to field labor or as a fraction of the value for the entire work. Each approach has advantages and disadvantages and the Project Manager must have the flexibility and foresight to tailor the monitoring system to the project needs.

Labor Related Work Unit Approach

In this approach, the estimated labor units (hours, days or weeks) are converted to work units and the one hundred percent completion mark will be a very specific number of W.U.'s that will change with every scope addition or deletion.

Proper application of this approach requires that the estimated field hours for every activity are known and that all have been estimated from the same data base (unit hours). It also requires that the hours reflect the real value and importance of the activities in terms of cost and/or risks involved.

Fraction Approach

Frequently it occurs that the impact of some of the major project components far exceeds the relative value based on hours alone. In those cases it is more appropriate to assign values based on a pondered assessment of:

- Dollar value.
- Importance to project.
- Risk exposure.
- Hours.
- Physical units (yd^3, ton, ft., etc).

Relative values are then assigned as decimal fractions of the total so that the completion target will always be 100% regardless of change. The example in Appendix F illustrates the use of the fraction approach.

Milestone Evaluation Method

Once all activities have been identified and assigned value, their progress must be monitored and measured in an objective, rational and consistent manner. This can be accomplished through the implementation of a milestone system that assigns value to each of the steps involved in the execution of a given activity and thus reduces the span of value susceptible to subjective evaluation.

Example - The installation of a pipeline involves seven steps valued as follows:

Activity	% Completion
Material receipt	5
Fabrication	25
Erection	20
Bolt up and weld	20
Install valves, hangers, trim	10
Hydrotest	10
Punch out	10
Total	100

The potential error in the evaluation of a line being erected should not be more than 5%.

Table 10.2 shows typical breakdowns for the most common field activities.

Schedule

A schedule is not required in order to gauge the percent completion at a given moment; all that is needed is a value system applied to the estimate items and a milestone system to gauge progress of each item. However, in order to determine whether the work is progressing as scheduled, a tracking curve is required.

The tracking curve is developed by superimposing (loading) the value of all the items included in the estimate on the project schedule. When the value system employed is based on hours only, the system also becomes, together with the daily force reports, an excellent way to monitor labor productivity.

Monitoring progress through the tracking curve alone could easily lead the project into trouble. While the curve shows that work is being accomplished at a certain rate, it does not indicate whether it is being done in the right areas. The schedule must be monitored to ascertain that the efforts are being applied in the right places and that the critical activities are being completed on time.

A project could be tracking beautifully until somebody finds that critical activity A, which requires one month and should have been completed two months ago, has not been started. Suddenly the project is three months behind schedule and there is not much that can be done about it.

In the ultimate monitoring system, every single estimate item would have a corresponding activity shown in the detailed schedule and a value based on hours. This level of precision can only be efficiently achieved with the aid of computer programs and would defeat the purpose of a quick checking system.

Table 10.2 Typical Construction Activity Breakdown

Activity	Percent Allocation
Concrete Foundations	
Excavate	15
Build and place form work	45
Place reinforcing steel	20
Pour and finish concrete	10
Strip forms and finish	10
Major Equipment	
Material receipt	5
Rough setting	65
Shim and plumb	25
Miscellaneous cleanup	5
Pumps and Compressors	
Material receipt	5
Rough setting	15
Level and grout	30
Align and couple	40
Lubricate and rotate	10
Piping	
Material receipt	5
Fabrication	25
Erection	20
Bolt up or weld	20
Install valves, hangers and trim	10
Hydrotest	10
Punch out	10
Equipment Insulation	
Receive and store	10
Install	70
Cover	20
Structural Steel	
Material receipt and storing	10
Erection	55
Plumb bolt up or weld	20
Trim and finalize	15
Buildings	
Structural frame	10
Roof	20
Walls	20
Doors and windows	10
Lighting	20
HVAC	20
Finish	10

Table 10.2 (Continued)

Activity	Percent Allocation
Instrumentation – Calibrate and Install	
Material receipt	5
Calibrate	15
Install	70
Activate	10
Instrumentation – Wire and Loop Check	
Conduit	20
Pull wire	40
Connect	20
Loop check	20
Instrumentation – Air Hookup	
Install pipe	30
Support pipe	30
Hook up	20
Activate	20
Electrical Power Feeders	
Material receipt	5
Conduit Runs	25
Conjunction or outlet boxes	20
Cable pulling	20
Connections	25
Ring out	5
Lighting	
Material receipt	5
Transformer & panels	20
Conduit runs	25
Cable pulling	20
Connections	20
Fixtures	10
Piping Insulation	
Receive and store	10
Straight runs	30
Fitting	40
Cover	20

For the in-house system proposed here, the tracking curve is based on the MPS discussed in Chapter 4. If the valve system is based on work hours, it should duplicate the construction progress curve included with the firm execution plan.

When values have been assigned based on fractions it is advisable to develop the tracking curve on the same basis. In any case, it is very important to remember that keeping close to the tracking curve could be inconsequential if the critical activities are not being done on schedule.

Progress Computation

Once all activities have been identified and assigned value and a milestone system established, the computation of earned values and overall progress (percent completion) is a very simple operation.

- At a given moment, many activities have neither been started nor others been completed. These are credited with zero or one hundred percent of their assigned value.
- The majority of those in progress will be covered by the milestone system and can be credited accordingly.
- The remaining few must be eyeballed. Of these, many will be either in the early stages or essentially complete; eyeballing them should present no problem.
- The potential errors of eyeballing any remaining activities should have a minimum effect on the overall evaluation.

The sum of the earned values divided by the total assigned value is the percent completion.

10.10   In-House Engineering Progress Monitoring System

Detailed System

The home office man-hours estimating procedure in Chapter 13 does not provide the detail of breakdown required for progress monitoring. However, once the list of required drawings and specifications has been prepared by the contractor, this information can be used together with Tables 10.3, 10.4 and 10.5 to prepare a simple monitoring system that will enable the Project Manager to ascertain the validity of the contractor's reports.

Table 10.3 lists the approximate hours required to prepare various types of drawings and equipment specifications. Although it could be used to check contractor's estimated hours, its prime purpose is to assign relative value to the principal home office activities.

Table 10.4 shows the percentage completion that should be assigned to the various milestones during the execution of each activity.

Table 10.5 shows a suggested breakdown for the engineering and design activities in general.

Table 10.3 Typical Home Office Work-Hour Units

	Phase 0	Phase I	Detailed Eng by Contractor	Total
A. Major Drawings				
	Work-hours per "E" Size Drawing			
PFDs (15-20 Equip. items)	40	40	0	80
P&IDs (6-7 Equip. items)	0	200	240	440
Layouts & Arrangements	0	0	120	120
Piping Orthographics	0	0	200	200
Civil/Struct/Arch	0	0	120	120
Electrical	0	0	160	160
B. Equipment Specifications				
	Work-hours per Specification			
Tanks & Vessels	8	32	40	80
Columns	24	56	64	144
Reactors	16	48	64	128
Heat Exchangers	8	16	16	40
Condensers with Inerts	16	24	24	64
Pumps	8	8	16	32
Utility Compressors	4	12	16	32
Agitators	4	12	16	32
Centrifuges	4	20	24	48
Package Systems Refrigeration, DMW, Cooling Towers, Inert Gas Generation, Steam Ejectors, etc.	8	24	32	64
Dry Material Handling	4	12	16	32
Miscellaneous Equipment	4	12	16	32
C. Miscellaneous Documentation				
General Specifications	0	80	160	240
Project Schedule	0	0	160	160
Coordination Procedure	0	0	80	80

Table 10.4 Typical Completion Milestones by Specific Activities

Activity	Percentage
Specifications	
Complete draft	20
Write specification	70
Check specification	80
Issue for approval	85
Issue revisions as required	95-100
Flowsheet Drawings	
Complete design calculations	40
Complete drawing	60
Check drawing	70
Issue for approval	80
Issue for construction	95
Issue revisions as required	95-100
Civil Drawings	
Complete preliminary site plan (building locations and site elevation)	10
Issue preliminary site plan for approval as required	25
Complete design calculations	30
Complete drawing	80
Issue for approval	85
Issue for construction	95
Issue revisions as required	95-100
Concrete and Foundation Drawings	
Complete design calculations	25
Complete drawing	60
Check drawing	85
Issue for approval	90
Issue for construction	95
Issue revisions as required	95-100
Architectural Drawings	
Complete sketches and general arrangements (GAs)	15
Issue sketches and GAs for approval as required	25
Complete drawing	75
Check drawing	85
Issue for approval	90
Issue for construction	95
Issue revisions as required	95-100
Steel and Superstructure Drawings	
Complete design calculations	25
Complete drawing	60
Check drawing	85
Issue for approval	90
Issue for construction	95
Issue revisions as required	95-100

Table 10.4 (Continued)

Activity	Percentage
Mechanical General Arrangement Drawings	
Complete design calculations	15
Complete preliminary GA drawings	20
Issue preliminary GA drawings for approval	40
Complete drawing	65
Check drawing	75
Issue for approval	80
Issue for construction	95
Issue revisions as required	95-100
Mechanical and Piping Drawings	
Complete design calculations	10
Complete drawing	65
Check drawing	85
Issue for approval	90
Issue for construction	95
Issue revisions as required	95-100
Electrical Drawings	
Complete design calculations	15
Complete drawing	75
Check drawing	85
Issue for approval	90
Issue for construction	95
Issue revisions as required	95-100
Instrumentation Drawings	
Complete design calculations	50
Complete drawing	70
Check drawing	85
Issue for approval	90
Issue for construction	95
Issue revisons as required	95-100

Quick System

The Project Manager does not always have the time to prepare and implement a full blown system. In these cases, the following empirical correlation can be used between 15 and 80% completion.

% Engineering complete = Significant drawing credit/Drawings required

Where:

Significant Drawing Credit: 0.25 (Drawings being worked) + Drawings issued.

Table 10.5 Typical Completion Milestones by General Activities

Activity	Percentage
Engineering	
Design criteria and scope of work verified, reference documents secured, process data secured	0-15
Specification and datasheet complete	15-40
Specification and datasheet issued to client for approval	40-60
Client approved and issued for bids	60-70
Bids analyzed and Purchase Requisition issued	70-85
Vendor prints reviewed	85-95
Vendor prints incorporated (does not include as-built activity)	95-100
Design	
Design criteria and scope of work verified; reference documents secured, manufacturer's data secured	0-15
Design calculations, sketches, instructions, etc., completed; other disciplines coordinated; drawing ready for final drafting	15-40
Drafting completed and back-checked	40-70
Independent check of calculations and drawings	70-80
Fix-up (drafting) and issue to project: project comments incorporated, for approval issue to client	80-85
Client comments incorporated and issued AFC	85-95
Vendor prints incorporated (does not include as-built activity)	95-100

Drawings Being Worked are the number of drawings actually started but not issued for construction. Use only the number actually being worked.

Drawings Issued are the number of drawings currently issued for construction *or* the number at least 90% complete.

Drawings Required are the number of drawings the contractor now estimates will be required to complete all design work. At the beginning of a project, this number is normally lower than what is finally needed. This should not affect the accuracy of the results, for there is a built-in allowance for this.

Drawings such as piping isometrics, pipe hangers, pipe support locations, instrument schematics, demolition and standard details should usually not be included in the numbers just defined. This normally deletes all drawings requiring less than 80 man-hours. The status of such drawings is too volatile and will distort the overall engineering status.

After discounting such types, only 35 to 40% remain from the number of drawings the contractor estimates as being required.

Since the relation is just a simple means of gauging the percentage completed, results should not be considered as exact, but as an indication of the status of engineering. If only a few percentage points' difference exists between the results of this formula and that reported by the contractor, probably no action is warranted. However, if the difference is substantial (more than 4 or 5 percentage points), an investigation is indicated.

10.11   Cost and Schedule Forecasts

Making periodic cost and schedule forecasts is the responsibility of the Project Manager. Management expects, rightfully, that the accuracy of these forecasts increases as the project moves along.

Accurate forecasts require careful analysis of the project performance to date in order to:

-   Determine if trends are evolving.
-   Understand the causes of deviations from the baselines.
-   Determine if the same causes are likely to affect the remaining work.
-   Evaluate the effect of any corrective action taken since the previous forecast.

The cost forecasts must be based on the continuing comparison during detailed engineering and construction of the actual versus estimated performance of:

-   Equipment and instrument counts as the P&ID's are completed.
-   Take-off quantities as discrete portions of the detailed design are completed.
-   Equipment and materials prices as firm vendors' quotes are received.
-   Labor rates and work-hours as subcontracts are awarded.
-   Rate of change orders as the field work proceeds.
-   Changes in local environment as they occur:
    1. Labor unrest.
    2. Inclement weather.

All cost forecasts must include resolution and contingency allowances to reflect the current level project definition.

The progress monitoring system, outlined in Section 10.9, is an excellent basis for schedule forecasting. It provides information that, when used in conjunction with daily field force reports and spot activity checks, will allow the Project Manager to make accurate schedule forecasts. The procedure for forecasting final subcontract costs in Appendix G is also a good tool for predicting the final projects cost.

10.12   Checking Contractors' Schedule and Execution Plan

General

The project execution plan and master project schedule developed by the Owner's Project Manager during the initial project stages (Chapter 4) set the basis and execution logic for the entire project and, normally, are all that is needed to execute small projects. However, in major projects, the responsibility for the actual execution is eventually turned over to an EPC contractor who must assume the planning and scheduling functions and issue an expanded execution plan and MPS in details commensurate with the size and complexity of the project.

The Owner's Project Manager still has the overall project execution responsibility and must ascertain that the contractor's proposed plan and schedule reflect all the available data and that they incorporate any out-of-the-ordinary action that will be taken in the course of the project to achieve the stated objectives.

- The review of the contractor's execution plan and schedule is a very important milestone. It is the last real opportunity to make corrections in the execution plan and, if required, revise the schedule forecast so that the financial plans can be adjusted to meet the new scenario.

- The approved execution plan and schedule must remain unchanged and become the basis to evaluate project performance. When contractors say "I meet all my schedules", they sometimes mean "I met the **last revision** of all my schedules." Even if the working schedules must be revised as required to preserve their meaningfulness, contractors should never be allowed to change the original approved schedule.

Review Criteria

The following factors must be investigated and taken into consideration in the preparation of the execution plan and schedule; it is up to the Owner's Project Manager to confirm that the contractor has done it in a conscientious manner.

General

- The submittal and approval by the pertinent authorities of environmental, construction and other permits required to operate the facility.
- The engineering schedule must support the construction contracting plan.
- The construction schedule must support the required start-up sequence.
- The schedule must allow adequate lead and/or turn-around time for:
  1. Vendors' and subcontractors' bidding.
  2. Owner's reviews and approvals.
  3. Vendors' drawings submittals and approval.
  4. Equipment deliveries.
  5. Final checkout.
- The critical path and floats must be clearly identified.
- The schedule must be kept as simple and unsophisticated as required by the complexity of the project.
- The use of the major construction equipment must be shown in the schedule.

Construction Staffing

- Available construction and layout areas.
- Available labor pool.
- Qualification and size of local contractors.
- Expiration dates of local labor contracts.

Lost Time Allowances

- Weather.
- Evacuation alarms.

- Holidays.
- Plant interferences.

Project Control

- Major process and/or geographical areas must be shown as independent self-contained schedules.
- Individual activities must be comprehensive and quantifiable.
- Temporary facilities and layout areas must be arranged to optimize the flow of work during construction.
- Ideally, each schedule activity should reflect a cost item in the estimate.

Schedule Constrains

- Every schedule constraint and/or bottleneck must be clearly identified.
- Any specific activity directed to improving the schedule must be identified.

10.13   Avoiding/Correcting Frequent Problems

Project control is about anticipating, avoiding, recognizing and correcting problems. This section is intended as a tool to help project managers to do so. It discusses some of the most frequent problems encountered in project execution, as well as the warning signs, and suggested avoiding and/or correcting action.

In the Engineering Contractor Office

Problem	Incompatibility with existing plant equipment numbering system.
Correction	Early in the project the Project Manager must insure agreement between the contractor and plant operations.
Problem	Ineffective communications. Contractor's personnel trying to follow conflicting instructions from Owner's team.
Correction	Contractor's personnel must be given instructions to listen politely to the Owner's team but act only on instructions coming from their own Project Manager. The contractor must follow up all meetings and project related communications with a written confirmation.

Problem	Special unscheduled studies interfere with the execution of the basic scope of work.
Correction	Have the contractor assign different personnel to special studies. Never give the contractor a "blank check". Negotiate the hours required to do the work before hand.
Problem	Owner's information and/or approvals are delaying contractor's work.
Correction	Owner's project team must realize that the contract is a two-way street. If they cannot live up to the commitments, the Project Manager must request more internal resources or consider assigning some of their work to the contractor.
Problem	Poor coordination within the contractor's task force.
Correction	Require that contractor's full time task force members be located physically in the same area.
Problem	Frequent changes in contractor's personnel.
Correction	Exercise contractual right of previous approval of key personnel changes. Probe the administrative qualification of the contractor's project manager, his/her degree of freedom and authority to approve/reject all personnel assigned to the task force.
Problem	Continuous changes in indicated final hours and large jumps between short reporting periods as engineering nears completion.
Correction	Require that requests for changes be submitted on weekly basis; approve only those related to change in scope. Refuse to honor those intended to cover for contractors estimating errors and inefficiencies as well as those submitted after the fact.
Problem	Concern for the proper functioning of the contractor's project control efforts.
Correction	Periodic reviews of work progress with each discipline supervisor in the presence of the contractor's project manager and his/her immediate superior.

During Procurement

Problem	Questionable quality of purchased equipment and/or materials.
Correction	Vendors must be prequalified jointly by contractor's and Owner's specialists.

Problem	Qualified vendor's shop's workloads endanger schedule.
Correction	Hold pre-bid and pre-award meetings with vendors to develop reasonable schedules. Get firm commitments and spread work over several shops. Work together with contractor's project manager to develop expediting and inspection schedules for major equipment.

Problem	Delays in issuing formal purchase orders.
Correction	Use telex or fax for immediate confirmation and follow up with formal documentation.

Problem	Delays in vendor's drawings jeopardize schedule.
Correction	Review and approve drawings for critical equipment at vendor's shop.

During Construction

Problem	Due to schedule pressures, the technical section of the bid packages contains holds and/or incomplete areas.
Correction	Holds and incomplete areas must be clearly identified and quantified based on the best available information to secure uniform bids. A set of applicable unit prices must be requested as part of the bid to adjust total cost after completion of design. Hold pre-award meeting to remove as many holds as possible.

Problem	Inconsistent bids due to different understanding of bid package.
Correction	Hold a pre-bid conference at the job site and insist that all bidders walk through the site of the job. Issue minutes of conference and answer all further questions in writing with copy to all bidders.

Problem	Available qualified labor in the area is tight due to high construction activity.
Correction	Conduct an area labor survey, interrogate prospective local bidders and, if necessary, consider outside bidders who could bring a group of their regular workers.

Problem	Improper planning of work by the subcontractor impacts the overall schedule.
Correction	The construction manager must insist that every subcontractor submit its own labor-loaded schedule and check them for compatibility with each other and with the master project schedule.
	If so required, the construction manager must prepare, with input from the subcontractors, a new MPS coordinating all work to insure completion by the established date.
	Weekly coordination meetings must be held with all subcontractors to insure compliance.

Problem	Inability to monitor the subcontractor's work progress.
Correction	Demand that all subcontractors base their work breakdown on the direct costs breakdown of the control estimate.

10.14   Work Sampling Guidelines

Work sampling is a technique used to determine the level of activity of a construction workforce. Although it is a good productivity indicator it should not be confused with the more sophisticated techniques used by industrial engineers to measure productivity. Work sampling is a powerful tool for the Owner's Project Manager and should be used in all projects.

While the reason for work sampling is more obvious on reimbursable jobs, observations on lump sum jobs help in determining the efficiency of the contractor's workforce and can alert the Project Manager to potential problems in either cost or schedule. Factual data on performance provides a good base for denying claims for extras when poor field performance is the cause.

Basically, the observer counts the people working and the total people observed. Since the crafts inherently resent work sampling, observations should be made instantaneously and unobtrusively as the observer tours the work area. Only those people actually performing physical work, such as welding, hammering, sawing, bolting, etc., are counted as working. For the purpose of this measurement, personnel carrying hand tools, waiting at the stockroom, waiting for machines or equipment, etc., are not considered as working. Routes and times of day should be varied so that the crafts cannot establish a pattern. Performing other observations at the same time serves to disguise the work sampling impact. It is not necessary to observe the entire workforce for this measurement. Obviously the larger sample provides more meaningful data.

Each run is logged with the date, time, observer's name, number working, total observed, percent working and remarks. Remarks should include comments on the weather and any unusual working conditions. The accuracy increases with the number of observations and observers.

The ratio of personnel working to total personnel observed provides the data for record. Graphing the daily observations is useful in spotting prolonged trends away from normal. One- or two-day deviations are not meaningful; however, deviations in excess of a week should be investigated actively.

The results can be compared to the data presented below which is based on data from thousands of observations from numerous types of jobs.

Under 30%: Productivity low, corrective action should be initiated

30% to 40%: Barely acceptable, below norm and needs attention.

40% to 50%: Good productivity.

Over 50%: Excellent, better than normal.

There are numerous reasons for poor productivity. The most often blamed and most overused reason is a poor attitude. Poor attitude, however, can stem from poor planning by supervision, the lack of tools or material, abusive supervision, unresolved grievances and numerous other causes. In any event, the cause should be investigated and action taken to correct the problem.

Conversely, sustained, above-average productivity is worthy of investigation to ascertain what techniques were used to provide it for future use.

CHAPTER 11
CONTRACT ADMINISTRATION

11.1 Overview

The Project Manager's concern is not only the physical conduct of the work, but also the implementation of all the contractual conditions. The Project Manager is also the Contract Administrator. Contract administration requires a total understanding of the contract, especially those clauses that assert the Owner's rights of technical approvals and control of the purse strings. The Project Manager, as well as every member of the project team, must be thoroughly familiar with the contract.

> A GOOD CONTRACT, WELL-UNDERSTOOD BY THE PROJECT MANAGER, WILL MINIMIZE CONFLICTS AND BE A VERY EFFECTIVE CONTROL TOOL.

The Project Manager is not expected to do a detailed audit of all invoices and backup materials, but only to monitor the critical items and point out to the accountants those areas where close scrutiny is warranted. The detailed audit is the job of accounting. It behooves the Project Manager to sit down with the Contract Engineer, at the beginning of the contract work, to review the contract, to understand all the critical provisions and to summarize them for the benefit of the members of the project team.

11.2 Thoughts on Contract Administration

Dealing with contractors is not an easy proposition. The contractor is neither thy enemy nor is it thy brother. The contractor is a business associate who shares only some of the client's objectives. Any responsible contractor will share the quality

---

objective with the client and, unless otherwise proven, must be given full credit for that. Beyond quality, the objectives usually start diverging:

- The contractor wants to maximize its profits.
- The client wants to optimize the cost/schedule equation.

Both objectives are legitimate and the contract minimizes conflicts by establishing procedures and regulating the actions of both parties. For example:

- In a lump sum contract, the contractor could increase the profits by understaffing and avoiding overtime at the expense of the schedule. This, of course, would conflict with the client's interests. This predicament would be avoided by a contractual clause penalizing the contractor for schedule delays.
- Conversely, on a reimbursable-plus-percent-fee contract, the contractor could overstaff and/or pay overtime in the name of schedule compliance. The result would be an increase in contractor's cash flow and profits but the improvement in the schedule, if any, may not justify the extra cost to the client. However, this situation would be avoided with a clause requiring prior client's approval to premium pay and staffing the job beyond the approved plan.

Although a contract ultimately implies an adversarial relationship, that concept cannot be brought down to the day-to-day dealings, especially with the contractor's Project Manager. It is important that both project managers establish a good rapport; after all they have a mutual objective, a successful project. However, both project managers must remember that they also represent opposite sides of a contract.

11.3 The Project Manager as Contract Administrator

The prime objective of the Project Manager acting as Contract Administrator is to protect the Owner's best interests by insuring that the work is executed in full compliance with the contract. Specific items include:

- Ascertaining that the contractor fulfills the promises made in the pro-posal and/or during negotiations:
  1. Personnel assignments.
  2. Subcontractors.
  3. Control system.
  4. Timely reports and schedules.

       5. Manpower loading.

- Exercising Owner's rights
    1. Technical approvals.
    2. Approval of expenditures (in reimbursable contracts).
- Reviewing and approving changes.
- Insuring that all the Owner's contractual commitments are met:
    1. Timely supply of information.
    2. Timely approvals.
    3. Permits and licenses.
    4. Communications through authorized channels only.

All contractual communications especially those concerning deviations, even minor ones, from the contract must be documented through:

- Memos,
- Minutes of meetings or even . . .
- Notes in a personal diary.

If the contract goes sour, a good diary may prove priceless.

Most contractors will take the initiative in issuing minutes of meetings and documenting conversations and scope variations. When doing so, being human, they may tend to emphasize the client's shortcomings while downplaying, or even ignoring, their own. It is up to the Owner's Project Manager to carefully review all their communications to make sure that they truly reflect the essence of what was actually said. When in disagreement, the Project Manager must let the contractor know immediately and document the disagreement. There is no need to write a formal memo; the original document can be returned to the contractor with proper annotations with a copy to the file.

The recommended way to start any contractual work is by holding a kickoff meeting attended by both project managers, their supervisors and the Contract Engineer. This meeting should be dedicated to the review of the scope and contractual conditions to make sure that everybody has the same understanding

The concerns associated with lump sum contracts differ from those associated with reimbursable contracts. In the former, the emphasis is on making the contractor live up to promises made in the proposal; the client wants to get as much as possible for the money. In a reimbursable contract, the emphasis is on keeping the contractor from doing, and spending, more than required; the client wants to keep costs to a minimum. In both cases, safety and soundness in design and construction are the compelling concerns.

The following is a list of some of the steps that the Project Manager should take as part of the project administration function. Those marked with an asterisk apply to both lump sum and reimbursable contracts. The rest are intended mainly for reimbursable contracts.

* At the beginning of the project, sponsor a meeting with the Owner's auditor and the contractor's accountants to work out the details of the invoicing and payment procedures and required backup documentation.
* Refuse any extra charges that have not been approved previously.
* Insist on the timely issue of reports and schedules. If necessary, issue a formal complaint.
* "Tighten the screws" but be careful not to relieve the contractor of its contractual responsibilities.
- Before contract award, clarify the dividing line between reimbursable services and those covered by overhead and profit.
- Identify the specific reimbursable personnel authorized to charge to the project and automatically refuse charges from unauthorized personnel.
- Watch for reimbursable personnel doing non-reimbursable tasks.
- Review and pre-approve personnel assignments on a weekly basis; followup.
- Ascertain that travel and living expenses are approved at the proper level in the contractor organization, consistent with the travel policy included in the contract.
- Maintain and continuously update a record of all commitments in a manner consistent with the estimate.
* Review the first invoices very carefully. Make a case of any error or omission no matter how minor. This will set the tone of the work.
- Review all payments made by accounting for consistency with the logged commitments.
- Have an auditor come and audit the contractor's records at 30% completion and maybe again at 70%.

11.4 Typical Audit Exceptions

The following is a list of exceptions normally taken by auditors reviewing project documentation for compliance with company policy and contractual conditions. This list will help the Project Manager with the contract administration work.

Purchasing - Lump Sum Contracts

- No authorization on increase in contract value.
- Additional work orders were made without a change in the scope of work for the contract on a lump sum project.
- Insurance charges reimbursed contrary to contractual agreement.
- Insurance certificate not on file.
- Expired performance bond.
- Significant rework in the field of vendor equipment without corresponding backcharge.
- Purchasing function not separated from receiving.
- Non-approved subcontractors.
- Expired contractor insurance.
- Subcontracts/purchase orders not written.
- Purchase bids and/or quotes not obtained for purchases over the value established in the contract.
- Purchase requisitions not approved.
- Purchase requisitioner bypassed purchasing.
- Sole-source purchasing not justified or approved at proper level.

Reimbursable Labor

- Construction timesheets not approved by C.M.
- No approval for overtime hours.
- Hours paid exceed hours on clock card.
- Early clock-outs not penalized.
- Time clocks tampered with.
- Fictitious employees.
- Contractor's employees not identified to brass alley.

Material

- Invoices not matched to receiving reports.
- No physical verification of materials received.

- Outbound shipments to be returned are not documented and no follow-up for return.
- Prices not verified to buying agreement.

Equipment and Property

- Records of rental equipment receipt, operation and departures not maintained nor equipment billings verified.
- Owner-owned property not tagged.
- Surplus equipment not identified.
- Owner-owned construction tools and equipment not physically verified.

Site Security

- Open and unattended gate.
- No security check of vehicles entering or leaving the site.
- No gate log maintained.
- No use of material passes.

CHAPTER 12
COMMUNICATIONS

12.1 Criteria and Guidelines

The Project Manager does not operate in a vacuum; the Project Manager is constantly:

- Receiving information from peers and consultants.
- Giving instructions to subordinates and contractors.
- Informing management.

THE PROJECT MANAGER MUST COMMUNICATE!

Being a good communicator:

- Verbally.
- In writing.
- Listening.

will invariably enhance the Project Manager's performance.

Proper and timely documentation is a necessary complement to communications and goes hand in hand with good project management and control.

- It keeps management informed, allows it to provide timely input and avoid unpleasant surprises.
- It generates hard records that could be very valuable for future analyses and decisions.

- It may be a catalyst to promote constructive thinking in both the writer and the readers.

On the other hand, overdocumentation is cumbersome, will dilute the efforts of the project team and slow down the project.

Discretion must be exercised to avoid unnecessary paperwork and ensure that the extent and detail of the documentation are commensurate with the size and importance of the project.

Documentation, especially memos and reports, must be clear, concise and attractive in order to ensure the attention of the recipients. Long, rambling and cluttered memos are often left unread and, if read, they are frequently misinterpreted.

With the advent of word processing and electronic mail, engineers are doing more and more of their writing on their computers. Obviously, this is a timesaving practice, but often the resulting documents are not the most attractive and/or understandable and are not suitable for issue outside the limited project team circle. On certain documents, engineers should either let qualified secretarial help do the typing or they should take some basic training in letter writing and editing.

12.2 Documentation Checklist

The following list points out the areas that should be documented during the course of a project. In some cases, the responsibility falls on the Project Manager. In some cases, it falls on others. But in all cases, the Project Manager must insist that the documentation is generated.

- **Project Kickoff** - The first formal contact with the client must be documented in a memo spelling out the request, requiring CED participation and client's desired schedule. Immediate action planned by CED should also be mentioned.

- **Initial Plan of Action** - After the client's request has been studied, a memo must be issued to inform all of the plan of action and timetable proposed to achieve the objectives. If the desired objectives are unrealistic, this must be brought to the client's attention and a realistic plan proposed.

- **Design Criteria** - The established design criteria and any changes to them must be clearly and promptly documented to ensure that both business group and operations personnel, as well as the design team, are fully informed and timely updated.

- **Meetings** - All meetings affecting project execution, decisions, instruction, technical reviews and approval must be properly documented in the minutes of meetings.

- **Scope** - A thorough and clear scope of work is essential to the start of any meaningful design activity either in-house or contracted. When properly reviewed and documented, it will give both the business groups and operations the opportunity to provide timely comments.

- **Design Packages** - Design packages (Phase 0 and Phase I) are normally self-explanatory. However, it is advisable to issue them with a cover memo summarizing the basic process information and design criteria followed.

- **Changes** - After the project has been approved based on a given scope, all changes and cost variations must be documented promptly to keep management informed and allow it to exercise the ultimate project control.

- **Estimates** - All estimates must be thoroughly documented. Documentation must not be limited to a cost summary. It must include an analysis discussing the basis of the estimate, estimating methods, areas of high uncertainty, assumptions, qualifications and contingency criteria.

- **Coordination Procedures** - The names, specific responsibilities and inter-relation of the key members of the project team and their line supervisors must be documented early for each specific project. Limits of authorization must also be clearly defined.

- **Contractors' Evaluations** - When a contractor is evaluated either for screening purposes or a specific contract award, the information developed must be documented for future reference.

- **Contract Award** - Every contract award whether competitive or negotiated must be documented with a thorough justification memo analyzing the reason for the selection and comparing the relative merits of all contractors considered.

- **Contractors' Performance** - The performance of active contractors must be evaluated and documented periodically.

- **Cost Evaluations** - In addition to the cost reports normally issued by the contractors, independent cost evaluations must be done and documented periodically by the Project Manager. Those evaluations must include analysis of variations and new cost forecasts.

- **Monthly Progress Reports** - Monthly reports must be structured to inform two different levels of management, a summary section addressed to corporate and group management, and a detailed section addressed to CED and division management.

- **Monthly Cost Reports** - Monthly cost reports must provide a clear view of the projected costs compared with AFE estimates plus a brief explanation of the changes from the previous report.
- **Deviations from Standards** - Any deviation from Corporate standards and guidelines must be justified and documented.
- **Project Closeout** - The completion of any project must be documented. Documentation should include relevant cost data that would be useful for future projects.

CHAPTER 13
SEMI-DETAILED ESTIMATING SYSTEM

13.1 Procedure

General

The proposed semi-detailed estimating system is consistent with the guidelines presented in Section 5.4 and includes simple procedures and tools that will allow project managers to quickly and accurately:

- Develop order of magnitude and conceptual estimates from minimal information.
- Prepare, when necessary, semi-detailed estimates (for small projects) suitable for appropriation and/or further project control.
- Check estimates prepared by others.
- Analyze subcontractors' bids.
- Check the cost of proposed changes.

When applied to Phase I designs (complete P&ID's and arrangement drawings), the quality of the resulting estimates could be as good as the conventional detailed estimates normally prepared by contractors at much greater expense.

The system emphasizes:

- The maximum use of equipment counts, rather than factors and ratios, as the prime parameter to determine the cost of commodities.
- The use of easily quantifiable comprehensive units of discrete comprehensive scope.

The system offers the following advantages:

- It offers the flexibility to prepare estimates at almost any level of engineering detail.
- It provides sufficient details to permit scope and cost tracking during the preliminary design stage.
- The details provided can be used directly to develop a progress monitoring system.
- It includes rational guidelines for consistent application of resolution allowances and contingency.
- Most of the procedures could be easily programmed in PC's using simple commercially available programs.
- All the estimating units can easily be adapted to reflect project-specific materials costs and labor rates.

Order of Magnitude and Conceptual Estimates

An order of magnitude estimate can be prepared in a few minutes, with the aid of the factors included in Section 13.11 when only the equipment count is available. When the equipment list with brief descriptions is also available, a conceptual estimate broken down by disciplines can be developed in a couple of days for a plant with 50 to 100 equipment items.

NOTE: ALL THE COSTS PROVIDED ARE BASED ON MID-1994 AND A CE PLAN COST INDEX OF 365.

- **Equipment Costs** - Use Section 13.2. Add 5% for freight.
- **Equipment Erection** - Use 20% of equipment cost as per Section 13.2.
- **Instrumentation** - Based on equipment count as per guidelines in Section 13.7.
- **Electrical** - Based on motor count as per guidelines in Section 13.6.
- **Engineering** - Estimate based on equipment list as per Section 13.8.
- **Piping** - Use a combination of:
  1. Factors in Chapter 5.
  2. Conceptual shortcut in Section 13.11.
  3. Unit prices and unit hours in Section 13.4.

    4. Judgment.
  - **Structural Steel** - Use a combination of:
    1. Factors in Chapter 5.
    2. Judgment to conceptualize the volume of structure required to contain the equipment and the pipe racks.
    3. Section 13.3 to estimate weight and cost.
  - **Labor Costs** - Follow guidelines in Section 13.9.
  - **Other Direct Accounts** - Use factors in Chapter 5.
  - **Field Indirects** - Since all costs and factors are based on subcontractor rates, use 20% of labor and subcontract columns.
  - **Resolution Allowance/Contingency** - Follow guidelines in Section 13.10.

Semi-Detailed Estimate

A semi-detailed estimate could be conceptual, preliminary or definitive, depending on the quality of the information provided. In any case, the following information is required:

  - Equipment list with sizes and materials of construction.
  - P&ID's.
  - Plot plans and equipment arrangements showing at least the outline of the required supporting structures.
  - Area of buildings.

If plot plans and equipment arrangements are not available, the estimator should be able to develop preliminary ones with the aid of Section 3.4.
    The preparation of a semi-detailed estimate for a medium sized chemical plant with 50 to 100 pieces of equipment should take 5 to 7 working days, including 3-4 days to develop preliminary equipment arrangements and doing all calculations manually. The various accounts are estimated as follows:

  - **Equipment** - Most equipment items can be estimated using Section 13.2 plus 5% for freight. Vendors should be contacted for special equipment.
  - **Piping** - Use P&ID's, Plot Plans and arrangement drawings to develop take-offs as follows:
    a. Identify and account for all process lines in a given process area and estimate their length starting from the nearest point in the

interconnecting pipe rack. Add around 10% to the shortest route to allow for the fact that it is not always feasible and 20% to hot lines (over 250°F) to allow for expansion loops.

b. Identify and account for all lines in the interconnecting pipe racks including utility supplies from remote areas.

c. Make specific allowances for distribution systems within the process area for the various utilities: cooling water, high/low pressure steam/condensate, service water/air, nitrogen, instrument air, vent collection systems, etc. Lengths and sizes must be determined based on arrangement and flows. Allow at least 30 ft of pipe from distribution system to each individual user.

d. Count all valves in P&ID's, including control valves. Allow one valve for every utility subheader. Do not count vents, drains and instrument connections; they are built into the pipe unit costs.

NOTE: The control valve count is used only to determine labor costs. The cost of the valve is included with the instrumentation account.

e. Establish the required quantity for each of the miscellaneous piping items mentioned in Section 13.4.
   1. **Heat Tracing** - Linear feet of tracing.
   2. **Utility Stations** - One per every 5,000 ft^2 of process area or one per level.
   3. **Safety Stations** - As required (at least one per operating area).
   4. **Instrument Purges** - Per process requirement.
   5. **Equipment Drains** - One per vessel or tank.
   6. **Sump Ejector** - One per sump or diked area.

f. Estimate material costs and labor hours with the aid of Section 13.4.

- **Instrumentation** - Use P&ID's to count balloons, TDC connections and air hook-ups. Compare count with average given in Section 13.7 and use unit costs.

- **Electrical** - Count motors and other electrical loads and use procedure in Section 13.6 to estimate cost.

NOTE: For a preliminary estimate, the cost can be calculated in a few minutes with the aid of the charts provided. For definitive estimates intended for appropriation and/or progress monitoring, the cost should be developed with the semi-detailed procedure. Both approaches should result in approximately the same cost.

- **Site Preparation, Sewers and Fire Protection** - If details are available, make rough take-offs and use prices in Section 13.3. If not, make an allowance consistent with the factors in Chapter 5.

- **Concrete** - Use procedure in Section 13.3 to develop take-offs and cost. Break down as many items as information permits.
  1. Equipment - by item.
  2. Structure Foundations - per structure.
  3. Floor Slabs - per areas.
  4. Dikes - per dike.

- **Structural Steel** - Use procedure in Section 13.3 to develop take-offs and cost. Break down by structures and pipe racks and don't forget to identify and estimate miscellaneous service ladders and platforms related to specific equipment items.

- **Buildings** - Identify and estimate the shelter (roof and siding) requirements for process structures as well as the area of buildings included in the Scope of Work. Use unit prices in Section 13.3 to estimate costs.

- **Insulation** - The take-offs for piping insulation are developed as part of the piping account estimate. The take-offs for equipment insulation are developed by identifying and estimating the outside area of those items requiring insulation. The unit prices in Section 13.5 are used to estimate cost.

- **Painting** - Make allowance consistent with the factors in Chapter 5.

- **Labor Costs** - All accounts estimated on a subcontract basis include the cost of labor and must be listed in the estimate under the "Material and Labor Subcontract" column. The hours included with the estimating units are intended only to determine construction staffing for planning purposes. Other accounts are estimated on a material and labor hours basis.
  The estimated hours must be adjusted, if necessary, for productivity as per the guidelines in Section 13.9. The cost must be estimated on a subcontract basis following the guidelines in Section 13.9 and shown in the estimate under the "Labor" column.

- **Field Indirects** - If the Owner acts as construction manager, the Field Indirects should not exceed 10% of the total labor and subcontract columns. If construction management is done by a contractor, follow guidelines in 13.9.

- **Engineering** - Contractor's engineering hours must be estimated with the procedure in Section 13.8. The cost per hour including overheads and profit varies from $70/hr in the East Coast and West Coast to $45/hr in the Midwest and South.

NOTE: The proposed execution approach will influence the required engineering hours and must be taken into account when preparing the estimate, especially a definitive or appropriation type estimate. Refer to the discussion in Section 10.3.

- **CED Costs** - CED costs will vary with the execution approach. If the project is going to be managed by CED using a full-service contractor, the required man-hours can be estimated with the aid of Section 13.8.
- **Taxes** - Taxes will depend on the specific location.
- **Spare Parts** - An allowance of 10% of equipment cost should be more than adequate.
- **Start-Up** - An allowance of 7% of the "Labor" column plus the labor fraction of the subcontract column is usually adequate.
- **Escalation** - Apply current inflation rate to the mid-life of the project. Refer to Section 13.10.
- **Resolution Allowance/Contingency** - Follow guidelines in Section 13.10.

Figure 13.0 summarizes the scope and applications of the proposed semi-detailed estimating system.

13.2 Equipment Estimating Procedures

General

The recommended, and most accurate, method of pricing equipment is with the help of vendors or recent purchase orders. However, it frequently occurs that either the purpose of the estimate does not justify the effort or the time constraint does not permit it. In those cases, the following procedures will be of great value to the Project Manager and/or the Estimator.

In the absence of more precise information, they have proved to be very quick and reasonably accurate especially when applied to a substantial number of pieces of equipment such as the entire equipment account of a chemical process plant. They are also useful to determine whether the price quoted for a specific piece of equipment is reasonable.

These procedures are more suitable for detailed or semi-detailed estimates where the various commodities (civil, piping, electrical, etc.) are estimated individually or related to equipment count rather than to equipment cost. These procedures should not be used in factored estimates where an error in the equipment account would be greatly amplified by the time it reaches the bottom line.

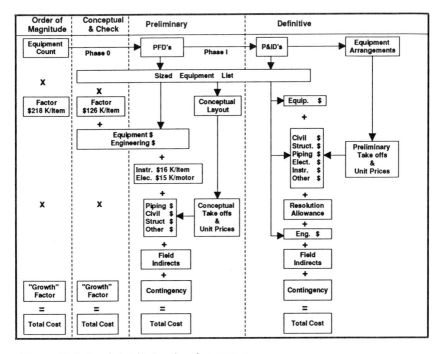

Figure 13.0 Semi-detailed estimating system.

Vessels

Scope

This procedure can be applied to all types of pressure and atmosphere tanks and vessels:

- Reactors, columns, storage tanks, process pots and pans.

It covers a variety of materials of construction:

- Carbon steel, 304 SS, 316 SS, Hastelloy C, glass lined, FRP Furan.

Weight Estimate

Basis of Procedure

The weight is determined with the aid of Figs. 13.1 or 13.2 and Fig. 13.3.

- Fig. 13.1 is used to estimate the thickness of steel vessels and tanks and is merely the graphical representation of the pressure vessel basic design equation.

    The estimated thickness is based on an allowable stress of 15000 psi and a joint efficiency of 0.85. It includes a corrosion allowance of one-eighth of one inch (1/8 in.).

    When the specified corrosion allowance is less than 1/8 in., the thickness must be adjusted accordingly. It must be noted that on flat bottom atmospheric storage tanks, the bottom and top thickness need not be more than 1/4 in.
- Fig. 13.2 was developed to estimate the thickness of FRP Furan tanks. It is based on 500 psi allowable stress and includes a 1/4-in. liner. The wall thickness in FRP tanks normally tapers down from bottom to top; for estimating purposes, the thickness should be calculated for the pressure at one-third of the tank height.
- Fig 13.3 was developed empirically to adjust the base level weight to allow for nozzles and supports.

Procedure

- Shell Area: $3.14 \times Dia \times Height$
- Flat Bottom Area: $(Dia)^2 \times .785$

- Cone Roof Area: (Dia)^2 x 1.1
- Approx. ASME Head Area   (Dia)^2
- Base Weight (C.S.):   Area x Unit Weight
  1. 1/6-in. plate   2.5 lb/ft^2
  2. 1/8-in. plate   5.0 lb/ft^2
  3. 1/4-in. plate   10.0 lb/ft^2
  4. 1/2-in. plate   20.0 lb/ft^2
  5. 1-in. plate   40.0 lb/ft^2
- Weight Corrections:
  1. Full vacuum service   x 1.10
  2. Hastelloy C   x 1.15
  3. Furan   x 0.20
  4. Baffles and Jackets   Calculate
  5. Nozzles and Supports   See Fig. 13.2

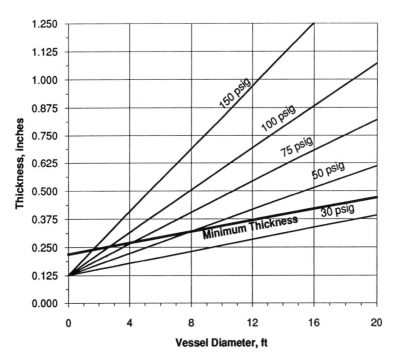

Figure 13.1 Metal vessel thickness.

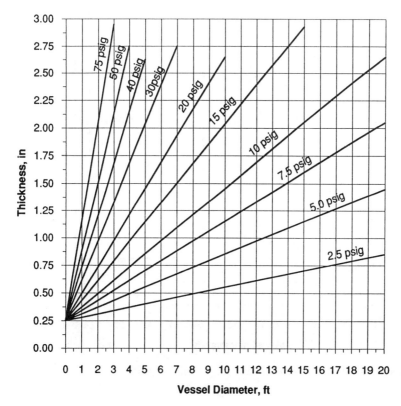

Figure 13.2 FRP (Furan) vessel thickness.

Cost Estimate

Step 1

Calculate cost of basic vessel from the following chart:

Weight lb		Cost $/lb	Weight lb	Cost $/lb
Up to	200	6.45	25000	1.95
	500	5.15	30000	1.75
	1000	4.50	40000	1.50
	2000	4.00	50000	1.35
	4000	3.55	70000	1.25
	6000	3.30	100000	1.10
	10000	2.90	150000	1.00
	15000	2.50	200000	0.95
	20000	2.50	200000	0.90

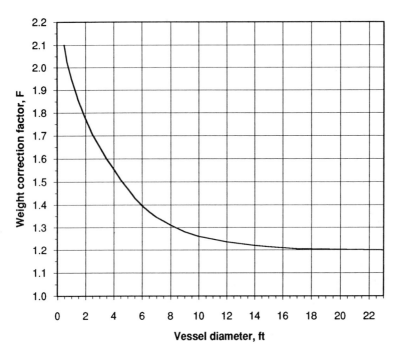

Figure 13.3 Vessel weight correction for nozzles and supports.

Basic Vessel Description

1. Carbon steel.
2. ASME construction (up to 75 psia).
3. Shop fabrication.
4. Simple design.

Step 2

Use correction factors to convert to actual conditions:

Material Correction	Multiplier
304L SS	2.20
316L SS	2.50
Glass lined	2.50
Hastelloy C	12.00
Furan	2.80

Fabrication Correction	Multiplier
Atm. shop fabricated	0.85
Atm. field fabricated, open top	0.90

Complexity Correction	Multiplier
Baffles, panel coils, pipe coils, external agitator supports	1.10 ea.
Full jacket	1.30
Reactors	1.50
Columns	1.50-2.00

Pumps

Fig. 13.4 was developed with cost data provided by a pump manufacturer and should cover the most frequently used pumps in chemical plants up to 100 hp. The base curves represent 316 SS horizontal ANSI pumps with TEFC motors and a single mechanical seal.

The cost of the basic pump (without motor) must be corrected as follows for the various cases covered by the procedure:

Material Correction	Multiplier
Ductile Iron	0.60
Hastelloy C	2.50

Pump Type Correction	Multiplier
API Pump	3.2
Dynamic Seal Pump	2.1
Magnetic Drive Pump	2.5

Double Seal Correction
Add $5,000 to account for:
1. Seal.
2. Seal pot and related instruments and interlocks.
3. Nitrogen supply.
4. Related piping and electrical wiring.

NOTE: Most of the costs associated with the double seal could be estimated with the piping, electrical and instrument accounts. However, to insure their inclusion in the estimate, they should be included together with the pumps.

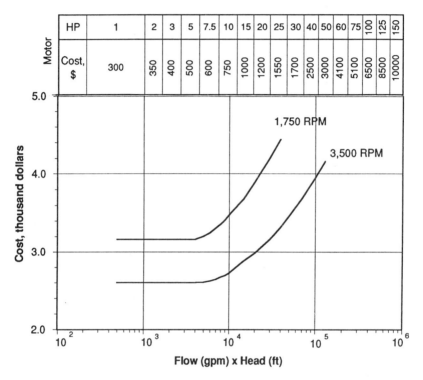

Figure 13.4 Centrifugal pumps cost.

Shell and Tube Heat Exchangers

Fig. 13.5 was developed with the aid of the B-JAC computer program to provide approximate cost estimates with minimal available information. The curve is based on the following parameters:

- Shell Material: Carbon Steel
- Tube Material: 304 SS
- Tube Arrangement: 1-in. Triangular
- Tube Diameter: 3/4-1 in., Welded
- Tube Sheet: Fixed

The following corrections are required for different construction materials:

Material	Shell	Tubes	Tube Ga.
Carbon Steel	0	-30%	16
304 SS	+14%	0	16
316 SS	+20%	+15%	16
Admiralty	+15%	+10%	16
Alloy 20	N.C.	+110%	18
Hastelloy C.	N.C.	+400%	18
Titanium	N.C.	+65%	20
Graphite	N.C.	+60%	-

N.C. - Not Considered

The cost of vacuum condensers would be 10% to 20% more than the corresponding heat exchanger depending on vacuum level.

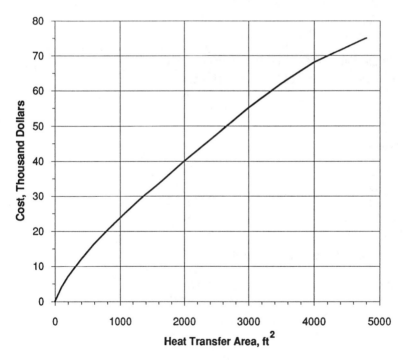

Figure 13.5 Shell and tube heat exchange cost.

Miscellaneous Equipment

The attached list provides costs of miscellaneous equipment frequently used in chemical plants. They were extracted from a widely used computerized estimating program:

Centrifugal Air Compressors - 50 to 100 psig
    1,000 cfm $115k to $120k
    5,000 cfm $260k to $276K

Reciprocal Air Compressors - 250 psig
    100 cfm $46k
    250 cfm $78k

Centrifugal Turbo Fans - 2 to 20 psig (carbon steel)
    500 cfm $21k
    5,000 cfm $63k to $88k

Rotary Blowers - 2 to 5 psig (carbon steel)
    500 cfm $16k
    1,000 cfm $19k

Centrifugal Fans - 6 to 15 in. H_2O (carbon steel)
    1,000 cfm $1k - $3k
    20,000 cfm $5k - $10k
    304 SS   x 1.5
    FRP   x 1.3

Air Dryers
    250 cfm $8k

Baghouses
    2,000 cfm $12k
    20,000 cfm $48k
    80,000 cfm $62k

Rotary Drum Blenders (carbon steel)
    50 C.F. $41k
    200 C.F. $124k
    304 SS   x 1.5
    316 SS   x 2.0

Fixed Propeller Agitators (316 SS)

5 hp	$12k
10 hp	$15k
20 hp	$21k
30 hp	$25k
50 hp	$36k
Carbon steel	x 0.9
Hastelloy C	x 2.5

Continuous Spray Dryers (carbon steel)

1,000 lb/hr Evap. Rate	$210k
5,000 lb/hr Evap. Rate	$325k
304 SS	x 1.8

Vacuum Tray Batch Dryers (carbon steel)

40 ft^2	$8k
200 ft^2	$22k
304 SS	x 1.5
316 SS	x 2.0

Batch Bottom Unloading Centrifuges (316 SS)

36 in. Dia.	$90k
56 in. Dia.	$140k

Continuous Solid Bowl Centrifuges (316 SS)

36 in. Dia. x 60 in.L	$280k
54 in. Dia. x 60 in.L	$310k

Open Belt Conveyors - 20 to 40 ft (carbon steel)

18 in. Wide	$21k to 25k

Screw Conveyors - 10 to 20 ft (carbon steel)

6 in. Dia.	$3k to $4k
9 in. Dia..	$4k to $5k
304 SS	x 1.5

Continuous Bucket Elevator - 30 to 50 ft (carbon steel)

10 in. Wide	$15k to $20k
16 in. Wide	$20k to $27k

Pneumatic Conveyor System (carbon steel)

100 ft x 4 in., 30 hp	$12k

Barometric Condensers (carbon steel)

200 gpm	$3k
2,000 gpm	$19k

Dowtherm Units - gas fired
    1.0 MM Btu/hr $56k
    10.0 MM Btu/hr $158k

Wiped Film Evaporator (316 SS)
    27 ft^2 $47k

4-Stage Steam Ejectors - 5.0 to 1.0 mm Hg
    20 lb/hr $11k to $21k
    50 lb/hr. $16k to $31k

Oil Seal Vacuum Pump (carbon steel)
    100 cfm $6k
    500 cfm $22k

Cartridge Filters (316 SS)
    50 gpm $3k
    200 gpm $4k
    500 gpm $7k

Pressure Leaf Filters (carbon steel)
    200 ft^2 $26k
    600 ft^2 $42k
    316 SS   x 2.0

Rotary Drum Filters (carbon steel)
    200 ft^2 $62k
    600 ft^2 $115k
    316 SS   x 2.0

Centrifugal Refrigeration Units - + 40 to -40°F
    75 tons $115k to $140k
    250 tons $315k to $380k

Gas Fired Package Boilers - 250 to 500 psig
    30,000 lb/hr. $160k to $175k
    100,000 lb/hr. $300k to $325k

Ion Exchange Water Treatment
    150 gpm $20k
    400 gpm $23k

Cooling Towers
    1,000 gpm $53k
    2,000 pgm $66k

Electric Generators
200 kW $47k
500 kW $118k

Stacks - 30 to 200 ft
24 in. Dia. $4k to $16k

Equipment Erection

The equipment erection hours are rarely more than 7-8% of the total construction hours, and errors in this account will have a very small impact on the total estimated costs. Estimating equipment erection costs as a percentage of the equipment cost is then an acceptable method.

IN MOST CASES, 20% OF THE EQUIPMENT COST WILL COVER THE COST OF ERECTION LABOR AND RELATED RENTAL EQUIPMENT.

This factor is to be applied only to the equipment listed in the estimate under the "Materials" column that requires minimum or no assembly by the regular construction forces. Equipment that requires field assembly by the vendor, such as large storage tanks, large boilers, materials handling systems and other complicated packages, should be listed in the estimate under the "Subcontract" column. The cost includes both materials and field costs.

The approximate field man-hours required for field manpower planning can be estimated as follows:

Erection Cost ÷ Loaded Labor Rate

Plus

(Subcontracts x 0.4) ÷ Loaded Labor Rate

Refer to Section 13.9 for applicable rates.

Knowing the total equipment erection hours is not sufficient for accurate construction progress monitoring; the total must be broken down into discrete portions related to specific equipment items. Table 13.1 provides guidelines for the allocation of hours to each equipment item.

Table 13.1 Typical Equipment Erection Work-Hours

Equipment	Unit	W-H
Pumps & drives		
< 1 hp	Ea	20
1-10 hp	Ea	40
15-25 hp	Ea	60
30-60 hp	Ea	80
70-125 hp	Ea	100
150 & over	Ea	120
Process vessels (< 12 ft dia.)		
< 100 gal	Ea	80
100-1,000 gal	Ea	100
1,100-5,000 gal	Ea	120
5,100-20,000 gal	Ea	160
Field fabricated vessels/tanks (> 12 ft dia.)		
< 20,000 gal	Ea	240
21,000-50,000 gal	Ea	480
51,000-100,000 gal	Ea	800
101,000-500,000 gal	Ea	1,200
501,000-1,000,000 gal	Ea	1,600
Compressors & refrigeration units		
< 50 hp	Ea	80
50-100 hp	Ea	100
100-300 hp	Ea	480
400-600 hp	Ea	880
700-1,000 hp	Ea	1,600
Agitators		
< 10 hp	Ea	30
15-50 hp	Ea	60
60-100 hp	Ea	100

Table 13.1 (Continued)

	Unit	W-H
Reactors		
< 500 gal	Ea	100
750	Ea	120
1,000	Ea	160
2,000	Ea	200
5,000	Ea	240
Shell & tube heat exchangers		
Up to 100 ft^2	Ea	40
200 ft^2	Ea	60
500 ft^2	Ea	80
1,000 ft^2	Ea	120
2,000 ft^2	Ea	160
Columns (typical)		
Up to 2ft Dia. x 30ft height	Ea	100
4ft Dia. x 50ft height	Ea	160
6ft Dia. x 80ft height	Ea	240
Columns internals		
2ft–5ft Dia. tanks	Ea	40
6ft–8ft Dia. tanks	Ea	60
Demister	Ea	20
Material handling equipment		
Field work-hours will vary greatly depending on the extent of field assembly required.		
Check with vendor.		

13.3   Civil Work Estimating Procedures

Concrete Work

This is not a foundation design procedure.   It is only a tool for preparing quick semi-detailed cost estimates and/or checking contractor estimates.   It is also a good tool for discussing and evaluating design alternatives and field change orders. They can be used either to estimate work done on a direct hire basis (materials plus labor) or work done on a subcontracted basis.

Unit Costs

Table 13.2 includes unit prices and work hours for the various types of concrete normally found in chemical projects.   They can be used to estimate work done on a subcontracted basis.

The units costs are comprehensive and include a pro rata of all related components and operations, from excavation to form stripping, clean-up and grouting.   They are consistent with costs obtained from a large civil contractor in the Northeast and based on loaded labor of approximately \$40/hr.

Table 13.2 Comprehensive Concrete Unit Costs

Description	Labor, W-Hr [1]	Subcontract, \$/CY
Ground slabs and area paving	6	300
Elevated slabs with metal deck	10	450
Equipment foundation, pile caps, grade beams	12	550
Structures & building footings, dike walls, sumps & pits	20	900
Structural concrete, columns & elevated beams	30	1,200
Average all types	12	550

Includes excavation, back fill, rebar, forming, stripping, finish, grouting & craft foremen.
Does not include dewatering, purchased backfill, underground obstructions, special concrete treating or curing.
[1]   For planning purposes only. Cost of labor is included in subcontract cost.

Labor hours represent the average of several contractors and published data and are based on "Gulf Coast" productivity.

The cost can be easily corrected to specific locations by adjusting the hours to reflect local productivity and the hourly rate to reflect labor rates and contractors' mark-ups.

Table 13.3 lists the unit hours for the individual operations normally involved in concrete work. These units could be very useful to discuss and negotiate field extras.

Material Take-Offs

Table 13.4 reflects a very rough design based on conservative assumptions and is used to approximate the concrete volume related to steel structures and pipe racks.

Fig. 13.6 Relates motor horse power to cubic yards of concrete and is used to estimate the approximate concrete volume for pumps. It assumes that the pumps will be set on individual foundations 4 ft below ground level.

Table 13.3 Miscellaneous Concrete Unit Work-Hours

Description	Unit	W-Hr	Description	Unit	W-Hr
Hand excavation one lift med. soil	yd^3	1.60	Rebar installation only [3]	lb	0.03
Mach. excavation (Typical) [1]	yd^3	0.10	Pour concrete (Average)	yd^3	2.50
Mach./hand excavation (Typical) [2]	yd^3	0.30	Concrete finish	ft^2	0.10
Back fill & compact (Manual)	yd^3	0.80	Anchor bolts & imbeded items	lb	0.10
Form work Fab./Install/Strip	ft^2	0.30	Grout	ft^2	0.30

[1] Backhoe med. soil.
[2] Mixed unit used by some contractors.
[3] Fabrication offsite.

Fig.13.7 relates motor horse power to cubic yards of concrete for both reciprocal and centrifugal compressors. It is based on the rule of thumb that foundation weight should be 5 times the compressor weight

Fig. 13.8 relates tank diameter to cubic yards of concrete for flat bottom tanks. It includes both concrete pads on top of existing slabs and independent ring foundations four feet deep.

Table 13.5 is used to estimate foundation volumes for tall tanks and towers. They have been derived from published information (Bauman, HC, *Fundamentals of Cost Engineering in the Chemical Industry*, Reinhold Publishing Co., 1964) and corrected for 125 mph winds.

Table 13.4 Comprehensive Concrete Take-Off Units

Description	CY Concrete per 1000 ft^3 Structure		CY Concrete per 100 ft^2 Floor Area		
	Foundation		Elev. Slabs 10 inch	Floor Slabs 6 inch	Tie Beams
	Open Structure	Sheltered Structure			
Heavy structures vibrating load	1.7	2.0	3.0	2.0	1.5
Heavy structures minimum vibration	1.5	1.8	3.0	2.0	1.5
Average structures	1.3	1.6	3.0	2.0	1.2
Light structures	1.1	1.4	3.0	2.0	1.0
Service structures	0.7	0.9	–	–	1.0
One level pipe racks	0.6	–	–	–	–
Two level pipe racks	0.8	–	–	–	–
Three level pipe racks	1.0	–	–	–	–

Dike walls 4 ft high	0.50 CY/ft
Dike walls 1 ft additional height	0.05 CY/ft

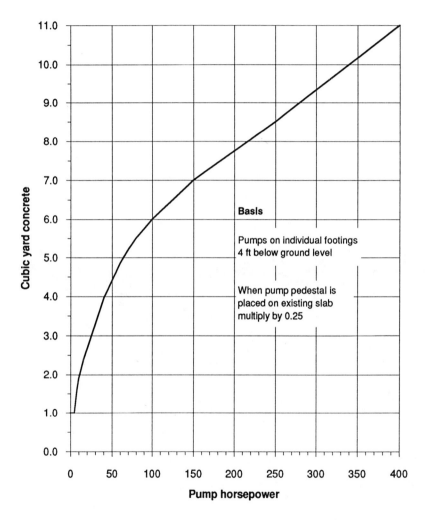

Figure 13.6 Pump foundation.

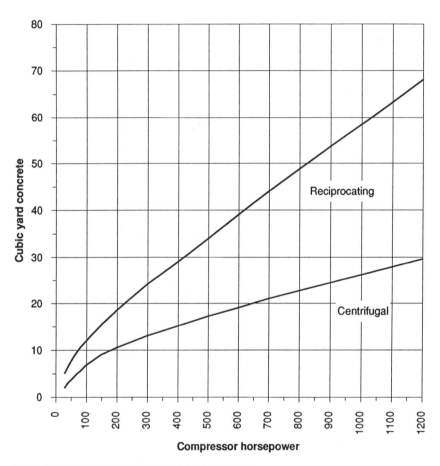

Basis: Weight of concrete is 5 times weight of compressor.

Figure 13.7 Compressor foundation.

Table 13.5 Columns Foundations

W/D	Cubic Yards			
	L/D = 3-8	L/D = 10	L/D = 15	L/D = 20
W based on empty vessel				
1	–	5	5	5
2	–	12	10	8
3	–	28	25	18
4	–	52	47	30
5	–	90	75	50
6	–	140	110	70
7	–	210	160	100
8	–	–	220	145
W based on test weight full of water				
10	10	–	–	–
20	28	–	–	–
30	45	–	–	–
40	65	–	–	–
50	85	–	–	–
60	105	–	–	–
70	135	–	–	–

Basis: Soil bearing: 2,000 lb/ft^2
Wind velocity: 125 mph

W - Weight, thousand pound.
D - Diameter, ft.
L - Height, ft.

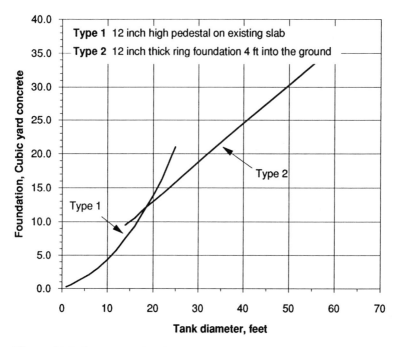

Figure 13.8 Concrete estimating system tank foundations.

Structural Steel

This procedure is used to estimate the weight, fabricated cost and erection hours of the various types of steel structures required in a chemical plant. It is a very good tool for checking contractors' take-offs and estimates as well as discussing and evaluating field extras.

Bases for Estimates

> **Material costs:** Fabricated and delivered to the job site, including one primer and two finish coats of shop applied paint.

> **Erection hours:** Represent the average of the hours used by several contractors and are based on competitive lump sum work basis.

> **Comprehensive unit weights (lb/ft^3):** Represent pondered averages derived from several actual cases.

Comprehensive unit costs: Developed from typical mixes of structural components in complex structures and the basic unit costs in Table 13.6.

Miscellaneous unit weights and costs: From published data and current costs.

Application

Complex structures: Calculate the total volume, including stairwells, to the highest platform and use the applicable units from Table 13.7.

NOTE: Do not include steel supporting roof and siding.

Table 13.6 Structural Steel Basic Units

Unit	Description	Cost, $ Per Ton	Cost, $ Other	Labor, WH Per Ton	Labor, WH Other
1	Heavy Steel, > 40 lb/ft	1,490	–	8	–
2	Medium Steel, 21 - 40 lb/ft	1,640	–	12	–
3	Light Steel, 11 - 20 lb/ft	2,050	–	25	–
4	Light steel, 0 - 11 lb/ft	2,560	–	40	–
5	Stairs - 2' - 6" wide w.o. handrailing	2,610	93.00/ft	27	1.00/ft
6	Handrailings with toe plate				
	Straight	3,900	25.60/ft	31	0.20/ft
	Circular	4,920	31.80/ft	31	0.20/ft
7	Ladders with cage	7,300	73.00/ft	44	0.45/ft
8	Galvanized floor grating 1 1/4 x 3/16	1,790	$7.80/ft^2$	32	$0.14/ft^2$
9	Miscellaneous platforms around tank, vessels, towers, etc. w.o. grating	4,000	$40.00/ft^2$	64	$0.65/ft^2$
10	Galvanized floor grating for circular platforms	3,790	$16.50/ft^2$	41	$0.18/ft^2$
11	Shop applied paint system included in unit prices	430	–	–	–

Table 13.7 Comprehensive Structural Steel Unit Costs

Unit	Description	Weight, lb/ft ^{3}	Cost, $/Ton	Erect. Labor, WH/Ton
1	Heavy equipment structures with vibrating loads	3.30	2,230	16
2	Heavy equipment structures with minimal vibration	3.00	2,280	18
3	Average equipment structures	2.60	2,460	21
4	Light equipment structures - some equipment on individual foundations	2.10	2,540	23
5	Service structures - Most equipment on individual foundations	1.50	3,310	32
6	1-Tier pipe racks	1.00	1,850	15
7	2-Tier pipe racks	1.20	1,850	19
8	3-Tier pipe racks	1.40	1,850	19
9	Shelter structures - No equipment - No roof or siding	1.40	1,850	19
10	Purlin system for corrugated roof and siding	3.00 per ft ^{2}	1,770	13

Units 1 through 5 include handrails, stairs and ladders. Do not include floor grating since in some cases floors may be concrete.
All costs include one primer and two finish paint coats applied at the shop.

Table 13.8 Miscellaneous Structural Steel Unit Weights

Description	Weight
1 1/4 x 3/16 bar type grating	9.7 lb/ft ^{2}
1 x 3/16 bar type grating	8.0 lb/ft ^{2}
Ladder 1'-6" w.o. cage	10.0 lb/ft
Ladder 1'-6" with cage	20.0 lb/ft
2" L handrail with toe plate	14.0 lb/ft
1 1/2" pipe handrail with toe plate	13.0 lb/ft
Stairways - 2'-6" wide w.o. handrails	71.0 lb/ft
Miscellaneous platforms around tanks & columns w.o. railing & grating	20.0 lb/ft ^{2}

Roof and siding: Calculate the volume of the structure supporting the roof and apply Unit 9 in Table 13.7. Calculate roof and wall area and apply Unit 10 in Table 13.7.

Pipe racks: Calculate volume and use applicable unit from Table 13.7.

Floor grating: Calculate area and apply Unit 8 in Table 13.6.

Miscellaneous: Apply units in Tables 13.6 and 13.8 as required.

NOTE: All inclusive (loaded) labor rates can be developed for different areas as follows:

Iron worker journeyman base rate	1.00
Direct supervision	0.15
Indirects - PAC's, small tools, overhead and profit	0.95
Multiplier	2.10

An additional $5/work-hour is suggested to cover major construction equipment.

Miscellaneous Civil Work

The following unit costs provide ballpark estimates for civil work frequently encountered in chemical plant projects. Some have been obtained from contractors, others from published cost data, and the rest reflect the author's actual past experience.

Site Work

- Site clearing	$4,200/acre
- Excavation and disposal within 10 miles	$16/y^3
- Compacted backfill including Materials	$26/y^3
- Asphalt roads (6 inch)	$21/y^2
- 8-ft Link fence	$26/ft
- 50-ft Wood piles	1,100 each
- Additional	$11/ft
- Metal sheet piling	$32/ft^2
- Railroad siding (including ballast)	$65/ft
- Railroad switches	$55,000 each

- Culverts (without excavation)
  1. 12-inch $15/ft
  2. 18-inch $21/ft
  3. 24-inch $32/ft
  4. 36-inch $75/ft
- U.G. PVC sewers (installed)
  1. 2-inch $21/ft
  2. 3-inch $23/ft
  3. 4-inch $26/ft
  4. 6-inch $32/ft
  5. 8-inch $38/ft
  6. 10-inch $45/ft

Fire Protection Systems

The following units can be used to develop ballpark estimates for the different components of fire protection systems.   They include excavation and backfill. When used to estimate grass roots plants, the estimated total should be checked with the factors in Chapter 5.

- U.G. water main
  1. 6-inch $42/ft
  2. 8-inch $47/ft
  3. 10-inch $55/ft
  4. 12-inch $64/ft
- Post indicator valves
  1. 6-inch $2,500 each
  2. 8-inch $2,700 each
  3. 10-inch $3,200 each
  4. 12-inch $3,800 each
- 6-inch hydrant including post indicator valve $6,000 each
  monitor and hoses
- Deluge sprinkler systems
  1. Extra hazard area (flammable service) $6.00/ft^2 floor area
  2. Regular hazard areas $5.00/ft^2 floor area
- Additional cost items

Insurance companies usually require that a highly reliable secondary water source be provided in-house. This could be in the form of an elevated water tank or a ground level tank or reservoir with a diesel operated pump. The capacity of the reservoir should be at least two hours of the system design flow capacity. The cost of these items can be estimated with the equipment estimating procedure and included with the equipment account.

Buildings

- A small or medium sized grass roots plant will require at least one building including:

	Approximate Area
Transformer switchgear room	600 ft^2
Control room	1,000 ft^2
MCC room	600 ft^2
2-3 production offices	400 ft^2
1 field laboratory	120 ft^2
1 lunch room	240 ft^2
1 change room	400 ft^2
Utility space	240 ft^2
Total	**3,000 ft^2**

Total Installed Cost (TIC)   Approx. $400k

- The following units can be used to develop ballpark estimates of different types of buildings frequently associated with chemical plants:

	TIC/ft^2 ($)
- Control room including computer floor 20 ft x 40 ft minimum	160
- MCC room including air conditioning 200 ft^2 minimum Plus 6 ft^2 per motor x 1.5	85
- Transformer switchgear room 20 ft x 30 ft minimum	85
- Miscellaneous buildings, maintenance shops, warehouse, guard house	65
- Front office	180
- Field office	85
- Laboratory	160
- Process buildings	130

Other Costs

- 8-inch concrete block wall including $8.00/ft²
  doors, windows, trims
- Corrugated 20-Ga. galvanized steel $4.50/ft²
  siding including doors, windows, trims
- 2-inch thick galvanized sandwich panel $8.00/ft²
  siding including doors, windows, trim
  but not the steel frame
- Pre-engineered metal buildings $30/ft²
  including foundation, insulation,
  doors, trim and lighting
- Uninterruptible power supply unit to $50,000 each
  be added to control room cost

13.4 Piping Estimating

Comprehensive Unit Prices

Tables 13.9 through 13.12 contain comprehensive unit prices and work hours for the fabrication and erection of process and interconnecting piping systems in carbon steel, 304 stainless steel, 316 stainless steel and plastic-lined carbon steel pipe.

They are intended for estimating large volumes of mixed diameter piping of the type typically found in chemical process plants. They should not be used to estimate buildings' service piping nor to estimate one single line, even for a chemical plant. However, the basic unit prices and hours in Tables 13.15, 13.16 and 13.17 can be used to develop estimates for any specific case, provided detailed take offs are available.

These comprehensive units, based on the models illustrated in Tables 13.13 and 13.14, include all the costs related to the fabrication, erection and testing of piping systems, except the cost of valves, which are priced separately.

Materials

- Pipe.
- Fittings.
- Flanges.
- Vents, drains and instrument connection.
- Non-structural hangers and supports.

Table 13.9 Comprehensive Piping Estimating Units - Carbon Steel

Diam., in	Sched.	Piping - per foot						Valves - Each		
		Process areas			Interconnecting					
			Labor, Hrs.			Labor, Hrs.			Labor, Hrs.	
		Mat'l, $	Fab.	Erect.	Mat'l, $	Fab.	Erect.	Mat'l, $	Fab.	Erect.
Up to 1	80	10.60	–	0.63	–	–	–	–	–	1.00
1-1/2	80	13.00	–	0.77	5.80	–	0.37	–	–	1.40
2	40	13.00	0.71	0.45	6.60	0.12	0.40	90.	2.00	2.40
3	40	14.90	0.73	0.53	8.60	0.14	0.46	110.	2.60	3.10
4	40	18.30	0.78	0.64	11.20	0.16	0.56	140.	3.10	3.80
6	40	26.00	0.97	0.85	16.60	0.19	0.75	210.	4.40	4.80
8	40	35.30	1.05	1.11	23.50	0.22	0.89	370.	5.40	6.80
10	Std.	48.00	1.08	1.35	33.30	0.26	1.04	620.	6.60	9.40
12	Std.	61.00	1.09	1.62	46.00	0.27	1.22	900.	7.70	11.00

Material: A106/A53 C.S Rating: 150 lb
Connections: SWD up to 1-1/2 B.W/flg. over 1-1/2
Service: Process - Non corrosive hydrocarbons & inorg. solutions & utilities

The materials cost represent current market prices and are shown in Tables 13.15 and 13.16.

Labor

- Unloading, storing and bagging.
- Layout and fit up.
- Shop and field welds and joints.
- Valves handling and installation.
- Installation of vents, drains and instrument connections (trims).
- Installation of hangers and supports.
- Hydrotesting.

Table 13.10 Comprehensive Piping Estimating Units - 304 SS

Diam., in	Sched.	Piping - per foot						Valves - Each		
		Process areas			Interconnecting					
		Mat'l, $	Labor, Hrs.		Mat'l, $	Labor, Hrs.		Mat'l, $	Labor, Hrs.	
			Fab.	Erect.		Fab.	Erect.		Fab.	Erect.
Up to 1	10	23.00	0.92	0.30	–	–	–	84.	2.0	1.50
1-1/2	10	23.60	0.97	0.38	12.60	0.14	0.36	86.	2.3	1.80
2	10	26.00	1.00	0.49	14.60	0.16	0.47	96.	3.2	2.40
3	10	32.00	1.10	0.59	19.80	0.19	0.53	132.	4.1	3.10
4	10	39.00	1.22	0.71	25.50	0.22	0.66	172.	5.0	3.80
6	10	61.00	1.49	0.96	40.00	0.28	0.89	316.	7.2	4.80
8	10	90.00	1.73	1.31	61.00	0.36	1.10	556.	9.4	6.80
10	10	128.00	1.90	1.63	89.00	0.43	1.32	900.	11.9	9.40
12	10	158.00	1.98	1.98	117.00	0.51	1.57	1,320.	15.0	11.00

Material: 304 SS   Rating: 150 lb.
Connections: Up to 3/4 S.W. 1in & over B.W. & L.J. flg.
Service: Process - Corrosive hydrocarbons & inorg. solutions.

The unit hours were derived from the basic units in Table 13.17, which represents the pondered average of nine different sources, contractors as well as published data. They are based on direct hire (reimbursable) work. For work done through a competitive lump sum contract, the figures must be reduced by 15%.

As shown in Tables 13.13 and 13.14, the fitting density varies with line size for the process area piping. The densities reflect actual experience in several projects and coincide with those used by others.

In addition to preparing estimates, these units can be very useful for:

- Checking estimates.
- Analyzing lump sum bids.
- Discussing and evaluating field changes.
- Monitoring field progress.

NOTE OF CAUTION: When checking estimates, it must be remembered that on sizes 2 1/2 in. and above, the process area pipe is usually prefabricated away from the field and some estimates would show pipe fabricated off-site as material cost. In that case, the cost will include:

- Bare materials.
- Cost of procuring material.
- Labor costs at shop rate.
- Shop overhead and profits.

Table 13.11 Comprehensive Piping Estimating Units - 316 SS

Diam., in	Sched.	Process areas			Interconnecting			Valves - Each		
		Mat'l, $	Labor, Hrs.		Mat'l, $	Labor, Hrs.		Mat'l, $	Labor, Hrs.	
			Fab.	Erect.		Fab.	Erect.		Fab.	Erect.
Up to 1	40	26.60	0.92	0.30	–	–	–	92.	2.0	1.50
1-1/2	10	27.20	0.97	0.38	15.00	0.14	0.36	94.	2.3	1.80
2	10	30.50	1.00	0.49	18.10	0.16	0.47	106.	3.2	2.40
3	10	38.50	1.10	0.59	24.70	0.19	0.53	146.	4.1	3.10
4	10	46.00	1.22	0.71	30.00	0.22	0.66	190.	5.0	3.80
6	10	74.00	1.49	0.96	49.00	0.28	0.89	348.	7.2	4.80
8	10	109.00	1.73	1.31	77.00	0.36	1.10	612.	9.4	6.80
10	10	160.00	1.90	1.63	113.00	0.43	1.32	990.	11.9	9.40
12	10	196.00	1.98	1.98	147.00	0.51	1.57	1,450.	17.0	11.00

Material: 316 SS   Rating: 150 lb.
Connections: Up to 3/4 S.W. 1in & over B.W. & L.J. flg.
Service: Process - Corrosive hydrocarbons & inorg. solutions.

Table 13.12   Comprehensive Piping Estimating Units:
Saran / PPL / Kynar / Teflon Lined Carbon Steel

Dia., in	Materials - $ / ft				Erection labor, Hrs/ft
	Saran 60°C	Polypropylene 100°C	Kynar 140°C	Teflon 200°C	
Process Areas					
1	60.60	64.30	95.00	93.00	0.60
1-1/2	63.00	69.00	83.00	96.00	0.64
2	66.50	73.00	91.00	103.00	0.68
3	79.00	85.00	113.00	125.00	0.77
4	95.00	103.00	144.00	158.00	0.85
6	104.00	153.00	220.00	260.00	1.05
8	195.00	218.00	307.00	380.00	1.24
Interconnecting					
1	19.70	21.00	27.80	35.00	0.30
1-1/2	22.20	24.60	33.80	40.00	0.32
2	25.00	28.30	39.20	46.00	0.37
3	33.30	37.70	54.60	62.00	0.44
4	43.30	49.00	75.50	85.00	0.53
6	69.00	80.00	125.00	152.00	0.69
8	109.00	123.00	184.00	250.00	0.82

Material: Schd 40 lined C.S rating: 150 lb.
Connections: Ductile iron flgs. & fittings.
Service: Corrosive hazardous liquids compatible with liner composition
and temperature rating.

Should process and/or safety conditions dictate the use of forged steel
flanges and fittings cost of materials will increase by:
15%   For process areas
10%   For interconnecting

Table 13.13 Comprehensive Piping Units Model: Carbon Steel / 304SS / 316SS

| | | | | | Per 100 feet of piping | | | |
| | | | | | | Welds | | |
Dia., in	Fittings	Flanges	Trims	Hang. & Supp.	Shop	Field	Joints	Bolt ups
Process Areas								
Up to 1	30	4	3	15	70 [1]	6 [1]	70 [2]	4
1-1/2	27	4	6	10	64 [1]	6 [1]	64 [2]	4
2	24	4	6	10	52	6	–	4
3	20	4	6	8	44	6	–	4
4	18	4	6	8	40	6	–	4
6	16	4	6	8	36	6	–	4
8	14	4	6	8	32	8	–	4
10	12	4	6	8	28	8	–	4
12	10	4	6	8	24	8	–	4
Interconnecting								
All	6	2	2	8	6	8	–	2

[1] Stainless steel pipe only.　　　　[2] Carbon steel pipe only.

Table 13.14 Comprehensive Piping Units Model: Plastic Lined Carbon Steel

| | | Per 100 feet of piping | | | | |
| | | Flanges | | | | |
Dia, in	Fittings	Shop	Field	Trims		Bolt ups
Process Areas						
1	30	47	2	3	12	25
1-1/2	27	42	2	6	10	22
2	24	38	2	6	10	20
3	20	32	2	6	8	17
4	18	29	2	6	8	16
6	16	26	2	6	8	14
8	14	23	2	6	8	13
Interconnecting						
All	6	10	2	2	10	7

Table 13.15 Basic Piping Materials Unit Costs in Dollars
Carbon Steel/304 SS/316 SS

	Diam. in								
	Up to 1	1.5	2	3	4	6	8	10	12
Carbon Steel									
Pipe, ft	0.74	1.20	1.66	3.32	4.74	7.80	11.70	16.60	22.10
Fttg, Ea.	6.60	15.30	9.00	12.90	19.80	42.30	78.00	140.00	206.00
Flg, Ea	17.00	18.00	21.00	25.00	34.00	53.00	78.00	115.00	170.00
Trim, Ea	100.00	100.00	100.00	100.00	100.00	100.00	100.00	100.00	100.00
304 SS									
Pipe, ft	5.15	6.10	7.50	11.50	15.00	23.70	37.20	50.40	64.30
Fttg, Ea.	10.80	10.80	12.40	22.90	37.40	110.00	215.00	384.00	554.00
Flg, Ea	32.00	32.00	37.00	47.00	62.00	111.00	185.00	390.00	510.00
Trim, Ea	200.00	200.00	200.00	200.00	200.00	200.00	200.00	200.00	200.00
316 SS									
Pipe, ft	6.70	7.80	10.00	15.50	19.10	30.80	49.20	67.30	83.10
Fttg, Ea.	13.60	13.60	15.30	28.60	47.70	137.00	266.00	481.00	708.00
Flg, Ea	35.00	35.00	41.00	52.00	69.00	123.00	213.00	445.00	680.00
Trim, Ea	220.00	220.00	220.00	220.00	220.00	220.00	220.00	220.00	220.00

Fitting	2/3 90° L.R. Ell + 1/3 Tee.	**All flanges**	Include prorata of gaskets, bolts & nuts.
SS Flanges	CS Slip on & SS stub end.	**Trims**	Include 3/4 in coup. and 3/4 in plug valve.

Table 13.16 Basic Piping Materials Unit Costs in Dollars
Plastic Lined Carbon Steel

	Diam. in						
	Up to 1	1.5	2	3	4	6	8
Saran Lined C.S.							
Pipe, ft	3.70	5.00	5.60	9.30	13.50	24.00	42.50
Fttg, Ea.	70.00	78.00	92.00	134.00	174.00	327.00	516.00
Flg, Ea	23.00	25.00	29.00	42.00	58.00	89.00	141.00
Trim, Ea	390.00	410.00	420.00	450.00	490.00	580.00	750.00
PPL Lined C.S.							
Pipe, ft	4.30	6.20	7.40	12.30	17.90	33.30	53.00
Fttg, Ea.	75.00	88.00	107.00	138.00	184.00	332.00	543.00
Flg, Ea	25.00	27.00	31.00	45.00	62.00	96.00	153.00
Trim, Ea	400.00	430.00	440.00	470.00	510.00	610.00	790.00
Kynar Lined C.S.							
Pipe, ft	9.50	13.50	15.50	24.50	36.00	64.50	94.50
Fttg, Ea.	76.00	93.00	112.00	173.00	216.00	465.00	795.00
Flg, Ea	31.00	33.00	40.00	56.00	99.00	120.00	167.00
Trim, Ea	430.00	490.00	500.00	540.00	590.00	720.00	900.00
Teflon Lined C.S.							
Pipe, ft	14.00	17.20	20.80	30.00	43.20	84.40	151.60
Fttg, Ea.	118.00	124.00	142.00	204.00	288.00	594.00	860.00
Flg, Ea	33.00	33.00	34.00	40.00	57.00	80.00	124.00
Trim, Ea	430.00	490.00	500.00	540.00	590.00	720.00	900.00

Table 13.17 Basic Piping Labor Units Based on Direct Hire Construction [1]

	\multicolumn{10}{c}{Diam, in}								
	0.75	1	1.5	2	3	4	6	8	10	12
Handling & Fit-up, Hrs/100 ft[2]										
Steel Pipe										
Proc.	20	20	20	25	30	35	50	60	75	90
Inter.	–	15	15	18	20	25	35	40	45	50
Lined Pipe										
Proc.	–	22	23	27	32	35	50	60	–	–
Inter.	–	12	12	14	15	20	30	35	–	–
Plastic Pipe										
Proc.	12	12	14	18	20	25	35	40	45	50
Inter.	–	8	10	12	14	16	25	30	35	40
Joints, Hrs/Ea										
Screwed [3]										
Shop	0.30	0.35	0.45	0.55	–	–	–	–	–	–
Field	0.35	0.40	0.55	0.70	–	–	–	–	–	–
S.W. Metal										
Shop	0.45	0.50	0.65	1.05	–	–	–	–	–	–
Field	0.50	0.60	0.80	1.30	–	–	–	–	–	–
Plastic Bell Conn										
Shop	0.40	0.40	0.40	0.45	0.55	0.60	0.70	0.80	1.10	1.20
Field	0.50	0.50	0.50	0.55	0.70	0.75	0.90	1.00	1.35	1.50
Butt Welds, Hrs/Ea										
A53 Cs [4]										
Schd	80	80	80	40	40	40	40	40	40	40
Shop	0.3	0.8	1.0	1.2	1.5	1.8	2.6	3.2	3.8	4.5
Field	1.0	1.0	1.2	1.5	1.9	2.3	3.2	4.0	4.8	5.6
304/316 [5]										
Schd	40	10	10	10	10	10	10	10	10	10
Shop	1.1	1.2	1.4	1.8	2.4	3.0	4.2	5.6	7.1	8.7
Field	1.4	1.5	1.7	2.3	3.0	3.7	5.3	7.0	8.9	10.9
Misc., Hrs/Ea										
Valves [6]	0.4	0.4	0.5	0.8	1.2	1.5	2.0	3.0	4.0	5.0
Bolt ups	0.7	0.7	0.8	1.0	1.2	1.5	1.8	2.5	3.5	4.0
Hang. & Supp.	1.0	1.0	1.2	1.5	2.0	2.5	3.0	3.5	4.0	5.0

[1] For competitive lump sum construction multiply by 0.85.
[2] Includes receiving, storing, rigging, aligning, tack welding (when required) and testing of all pipe & fittings.
[3] Includes cutting & threading.
[4] For Cr-Moly Cs add 25%.
[5] Correction for different alloys: Alloy 20/Monel/Aluminum – add 5%.
     Nickel/Hasteloy – add 40%.
[6] Handling only: Installation of companion flanges and bolt ups must be added.

Miscellaneous Comprehensive Unit Prices

Some piping items will usually not be shown in the P&ID's and are often missed in the estimates. The items included in Table 13.18 should cover most of these cases. They should be addressed when checking an estimate or preparing one.

Heat Tracing - The heat tracing units, steam and electrical, were developed in the late 1980's with the help of construction contractors and have been escalated to mid-1994. The models used to prepare the unit prices are shown in Appendix H and can be used to adapt them to different circumstances.

The heat tracing costs could be reduced by 20-30% with judicious design and optimization of the length of pipe that can be traced with the maximum allowable tracer. However, no credit should be taken for optimization until detailed engineering is done.

Utility Stations - Utility stations (steam, air and water) with 50-ft hoses are normally required in process areas. A station should be provided for approximately every 5,000 ft^2 of process area or at least one per level of structure. This unit includes three 50-ft runs of 1-in. carbon steel pipe with one service valve each. The hoses, heat tracing and insulation are not included.

Safety Stations - Most process plants require safety showers and eye wash fountains strategically located in the process area. The number must be determined based on safety requirements and operating areas. This unit includes one safety shower, one eye wash fountain, 50 ft of 3/4-in. carbon steel pipe and two 3/4-in. valves. Steam tracing, insulation and lights are not included.

Instrument Purge - On occasion the process requires that some instrument be purged with either air or inert gas. This unit includes 50 ft of 1/2-in. carbon steel line and one valve.

Equipment Drains - Most equipment items have drains which must be run to nearby collecting systems. The size will vary with the size of the equipment, but 1 1/2-in. is a realistic average size. The units include one 1 1/2-in. valve and 20 ft of process type pipe.

Sump Ejector - Curbed areas and dikes are usually provided with a sump and a small steam ejector to empty it. This unit requires insulated steam piping, valve, ejector and condensate traps.

Steam Traps - Steam trap assemblies are usually shown in the P&ID's but no cost has been provided elsewhere. An allowance is included here.

Table 13.18 Miscellaneous Piping Estimating Units

Description	Unit	Mat'l, $	Labor, WH	S.C., $
A. Heat Tracing				
Electric				
Up to 4" pipe 1-Tracer				
Winterizing service	LF	13.50	0.16	22.30
250°F process service	LF	23.80	0.20	35.40
300°F process service	LF	30.10	0.16	44.70
6" to 10" Pipe 2-Tracers				
Winterizing service	LF	27.00	0.32	44.70
250°F process service	LF	47.60	0.40	70.90
300°F process service	LF	60.20	0.32	81.20
12" & Up Pipe 3-Tracers				
Winterizing service	LF	40.50	0.48	67.10
250°F process service	LF	71.40	0.60	106.30
300°F process service	LF	90.20	0.48	121.70
Steam				
Up to 4" pipe 1-Tracer				
Screwed & welded pipe	LF	22.70	0.80	63.40
Flanged pipe	LF	24.70	1.10	73.60
6" to 10" Pipe 2-Tracers				
Screwed & welded pipe	LF	45.30	1.60	126.70
Flanged pipe	LF	49.40	2.20	190.00
12" & Up Pipe 3-Tracers				
Screwed & welded pipe	LF	68.00	2.40	190.30
Flanged pipe	LF	74.20	3.30	221.00
B. Miscellaneous Composite Units				
Utility station water/air/steam	Ea	900.00	125	–
Safety station shower/eye wash	Ea	630.00	50	–
Instrument purge	Ea	220.00	25	–
C.S equip drain	Ea	260.00	10	–
304SS equip drain	Ea	860.00	15	–
Sump ejector	Ea	520.00	20	–
Steam traps	Ea	1,030.00	20	–

Utility Headers - Allowances should be made to run utility headers (service and instrument air, cooling water supply and return, steam, process water, potable water, etc.) through the longest axis of the process units. Size varies with the plant. Use the interconnecting piping units to estimate utility headers.

The instrument air distribution system (headers and subheaders) is considered part of the piping account up to the distribution points to the individual users. The lines to the individual instruments and control valves are considered part of the instrumentation account and the cost is included in the instrumentation estimate.

NOTE: Another item often overlooked is the piping, instrumentation and electrical costs associated with double seal pumps (seal oil pot, piping, pressure controls and interlocks). The cost is included in the pump estimated procedure.

Miscellaneous Valves Costs

Table 13.19 represents the average cost of valves purchased at different times escalated to mid-1994. These costs should only be used for conceptual or preliminary estimates when no prices are available. Since the cost variations from one manufacturer to another can be astronomical, definitive estimates should be based on vendors' quotes or, at least, on recent purchases.

13.5 Insulation Estimating

Tables 13.20 through 13.23 can be used either for semi-detailed or detailed estimating of piping and equipment, hot and cold insulation.

The basic piping insulation units in Table 13.20 and the equipment insulation units in Table 13.23 were provided by an Eastern Shore contractor and have been escalated to mid-1994 level.

The comprehensive piping units in Tables 13.21 and 13.22 were developed with the basic units in Table 13.20 and the fitting density used for the piping comprehensive units (Table 13.13) applying the linear feet-per-fitting equivalence normally used in the trade:

-	Fittings	
	1. Up to 2 in.	1.3 ft/ftg.
	2. 3 in.	2.0 ft/ftg.
	3. 4 in. and up.	3.0 ft/ftg.
-	Flanges	4.0 ft/ftg.

Table 13.19 Approximate Cost of Valves

Conn.	Diam., in	Ball	Plug	Check	Gate	Globe	Btly
				Carbon Steel			
SWD or SW							
	1	70	60	40	50	60	50
	1.5	80	80	70	90	110	60
Flg.							
	2	190	160	200	230	300	150
	3	340	290	290	320	470	190
	4	510	440	380	390	630	240
	6	920	780	550	650	930	380
	8	1,720	1,470	950	920	1,230	540
	10	2,800	2,390	1,450	1,450	2,250	720
	12	4,160	3,560	2,050	2,100	3,400	1,100
				316 Stainless Steel			
SWD or SW							
	1	140	130	200	210	260	100
	1.5	230	220	310	290	370	130
Flg.							
	2	430	370	380	420	480	320
	3	750	640	650	610	830	420
	4	1,130	970	950	800	1,240	520
	6	2,700	2,300	1,600	1,400	3,000	850
	8	5,000	4,300	3,100	2,700	5,700	1,200
	10	8,100	6,900	5,000	4,500	9,300	1,600
	12	11,400	10,200	7,500	6,900	14,100	2,000
				Teflon Lined Carbon Steel			
Flg.							
	1	120	110	70	90	110	80
	1.5	150	140	130	150	200	110
	2	340	290	350	430	550	260
	3	610	520	520	570	850	350
	4	930	800	690	710	1,160	420
	6	1,680	1,440	1,010	1,200	1,800	690
	8	3,200	2,700	1,740	1,700	2,500	990
	10	5,100	4,400	2,700	2,700	4,100	1,300
	12	7,600	6,500	3,800	3,800	6,200	1,600

Table 13.20 Basic piping insulation units

Diam., in	Dollars & work-hours per L.F.											
	1 in thk		1.5 in thk		2 in thk		2.5 in thk		3 in thk		4 in thk	
	Mat'l	Lab	Mat'l	Lab	Mat'l	Lab	Mat'l	Lab	Mat'l	Lab	Mat'l	Lab
Up to 1	2.40	0.20	4.15	0.21	6.50	0.23	7.40	0.26	9.90	0.31	16.50	0.38
1.5	2.65	0.20	4.70	0.21	7.20	0.23	8.20	0.26	10.60	0.31	17.20	0.38
2	2.85	0.21	5.10	0.22	7.50	0.24	8.40	0.27	11.00	0.32	17.30	0.39
3	3.55	0.22	5.80	0.23	8.70	0.25	10.30	0.28	12.60	0.34	19.40	0.41
4	4.60	0.23	6.60	0.24	10.00	0.26	11.50	0.30	14.70	0.36	21.10	0.44
6	5.60	0.25	7.60	0.27	11.60	0.31	15.30	0.35	17.90	0.42	25.50	0.50
8	8.00	0.29	9.50	0.32	14.40	0.36	18.20	0.42	22.10	0.48	28.20	0.58
10	9.40	0.34	11.90	0.37	17.20	0.42	21.80	0.49	25.50	0.55	34.20	0.68
12	10.70	0.39	13.00	0.43	18.90	0.50	23.80	0.57	28.60	0.66	37.70	0.80
14	12.40	0.45	15.00	0.50	21.40	0.57	26.90	0.66	32.50	0.76	40.90	0.93
16	14.70	0.52	16.90	0.58	23.50	0.66	29.80	0.75	36.80	0.87	45.30	1.05
18	15.80	0.57	19.20	0.66	25.60	0.75	32.50	0.85	40.00	0.98	49.20	1.18
20	17.60	0.64	20.40	0.75	28.30	0.84	35.40	0.96	43.20	1.10	52.60	1.32

Conversion Factors
Foam glass w SS jacket
    Mat'l x 1.30
    Lab x 1.35
Alum. jacket: Mat'l x 0.90
PVC jacket: Mat'l x 0.85

Material: Fiber glass & SS jacket
Service: Up to 400°F

All units are based on doing the work at an average height
of 10 to 30 feet above floor level.

Table 13.21 Comprehensive Piping Insulation Estimating Units: Hot Service

Diam., in	Thickness, in	Piping – per L.F.				Valves,	
		Process & utility areas		Interconnecting		Each	
		Lab, hr	S.C., $	Lab, hr	S.C., $	Lab, hr	S.C., $
Up to 1	2	0.36	25.60	0.27	18.90	0.30	22.60
1.5	2	0.36	26.40	0.27	19.70	0.50	32.50
2	2-1/2	0.41	30.00	0.32	23.20	2.60	143.00
3	3	0.53	42.00	0.43	34.00	3.40	274.00
4	3	0.61	50.60	0.48	35.60	4.10	339.00
6	3-1/2	0.76	66.30	0.61	53.30	5.10	455.00
8	4	0.92	83.20	0.76	62.40	6.40	590.00
10	4	1.03	95.30	0.90	82.50	7.50	766.00
12	4	1.17	104.00	1.06	94.30	8.80	800.00
14	4	1.33	114.00	1.23	106.00	10.10	893.00
16	4	1.47	125.00	1.39	118.00	11.60	1,020.00
18	4	1.62	136.00	1.56	130.00	13.00	1,110.00
20	4	1.81	148.00	1.74	142.00	14.40	1,205.00

Sub contract = Mat'l + work-hour @ $42/hr
Material: Fiber glass & S.S. jacket
Service: Up to 400°F

Table 13.22   Comprehensive Piping Insulation Estimating Units:
                          Cold Service

Diam., in	Thickness, in	Piping – per L.F.				Valves,	
		Process & utility areas		Interconnecting		Each	
		Lab, hr	S.C., $	Lab, hr	S.C., $	Lab, hr	S.C., $
1	2	0.49	34.20	0.36	25.00	0.40	29.80
1.5	2	0.49	35.30	0.36	26.00	0.63	44.10
2	2	0.49	35.40	0.38	27.40	3.50	256.00
3	2.5	0.59	45.70	0.48	37.10	4.60	363.00
4	2.5	0.69	54.40	0.53	41.60	5.50	448.00
6	2.5	0.75	64.10	0.62	52.30	6.80	603.00
8	2.5	0.90	75.20	0.75	62.80	8.60	780.00
10	2.5	1.00	85.10	0.87	74.00	11.10	932.00
12	2.5	1.12	89.20	1.01	83.30	11.50	1,060.00
14	3	1.47	122.00	1.35	113.00	15.60	1,130.00
16	3	1.64	136.00	1.55	128.00	15.70	1,350.00
18	3	1.81	147.00	1.74	142.00	17.60	1,470.00
20	3	2.03	112.00	1.96	157.00	19.40	1,590.00

Sub contract = Mat'l + work-hour @ $42/hr
Material: Fiber glass & SS jacket
Service: Down to 6°F

Table 13.23 Equipment Insulation Estimating Units

Insulation Material	Thk, in	Per S.F.	
		Lab, hr	S.C., $
	1-1/2	0.19	11.60
	2	0.19	12.80
Fiberglass with	2-1/2	0.20	14.60
S.S. jacket	3	0.10	16.20
	3-1/2	0.21	17.20
	4	0.24	19.60
	1-1/2	0.25	15.00
	2	0.25	16.50
Foam glass with	2-1/2	0.26	18.70
S.S. jacket	3	0.26	20.70
	3-1/2	0.28	22.30
	4	0.32	25.30

13.6 Electrical Work Estimating Procedure

Introduction

Accurate cost estimates for electrical work can be prepared after a reasonable amount of engineering has been completed. This method is time consuming and costly and, if the project is not approved, wasteful.

At the other extreme, electrical work could be estimated as a percent of the equipment costs. However, this method is not only very inaccurate (published information shows a range of 10% to 50% of equipment cost), but also does not provide a means of cost tracking or progress monitoring.

The cost of the electrical account is mainly related to motor count and average horsepower per motor rather than to equipment cost or total horsepower.

- The cost of connecting the motor of a $1,000 pump is the same as for the motor of a $50,000 agitator.
- The cost of connecting one 100-hp motor will probably be less than 20% of the cost of connecting ten 10-hp motors.
- The process areas requiring lighting will normally be directly proportional to the number of motors and average horsepower.

This procedure presents a quick and reasonably accurate system to estimate the electrical account at different levels of engineering design in sufficient detail for subsequent cost tracking and progress monitoring. It can be used for grass roots plants/units as well as for retrofit work.

Scope

The estimating units are intended in general to include all electrical materials and installation costs, except as noted below, for process and utility areas up to and including 480 v substations and associated switchgear.

- Instruments and DCS wiring are considered part of the instrumentation account and the pertinent estimating units have been included in the instrumentation estimating procedure.
- The cost of main substation, high-voltage distribution and yard lighting must be estimated separately.

The comprehensive materials and work-hours charts (Figures 13.9 and 13.10) were developed from a theoretical model of a plant with one hundred 460 v motors of different sizes, half of them with interlocks, including allowance for lighting, welding receptacles, 110 v outlets and grounding, directly proportional to the average hp/motor. The material costs and labor hours are based on the composite unit costs in Table 13.24.

The unit prices and unit hours represent composites of all materials and operations required for a finished product; i.e., the motor hook-up units include the breakers, push button, a pro rata of the MCC cabinet conduits or cable trays, power and control wiring, terminators, miscellaneous supports, unloading and storing materials, testing and commissioning, etc.

All units are based on Class I, Division 2 service. Occasionally, some sections of a plant are classified as Division 1. In those individual areas, the costs may increase by as much as 50%.

Application

Conceptual Estimate

A conceptual estimate can be developed in a few minutes from the preliminary equipment list as follows:

- Identify and count pumps, agitators, compressors, material handling equipment and other items requiring electric motors.
- Identify and count 460 v electric heaters; treat them as motors.

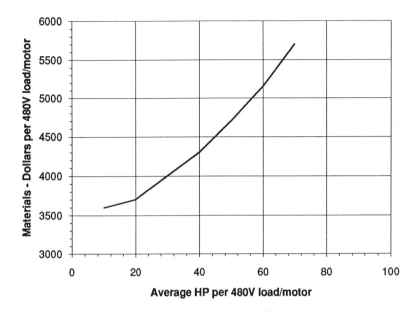

Figure 13.9 Cost of installation materials per 480V load/motor.

- Make allowance for multiple motors in packages, such as lubricating pumps.
- Make allowance for electric motors not included in equipment list, such as fans, HVAC equipment, motor-operated valves and doors, etc., using 480 v.
- Estimate cost by multiplying total motor count by $18,000; approximately $10,700 for materials; and 175 work-hours at $42/hr.

NOTE: This cost is valid only for plants averaging 20 hp/motor or less and does not include the cost of inactive areas lighting.

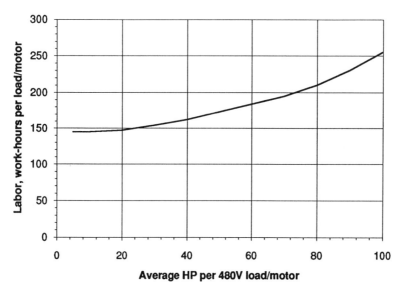

Includes:
- All 480V power wiring for both process and non-process service
- Process area lighting and misc. 110V loads
- Hard wire interlocks
- Grounding

Excludes:
- Main substation & high volt dist.
- 480V transformer & switch gear
- High voltage motors wiring
- Instrumentation wiring
- Electric heat tracing
- Non-process buildings lighting
- Yard lighting

Figure 13.10 Installation labor per 480V load/motor.

Preliminary Estimate

A preliminary estimate can be developed in less than two hours with a complete equipment list showing approximate motor horsepower:

- Count 460 v motors and electric loads as before.
- Identify motors operating above 460 v.
- Calculate total 460 v load and average horsepower per motor.
- Use Figures 13.9 and 13.10 to determine cost of 460 v and 120 v system and Table 13.24 to determine the cost of the substation (1,000 kva module), the large horsepower motors and the off-site lighting.

Table 13.24 Comprehensive Electrical Estimating Units

Volts	Description	Unit	Fixed Cost		Variable Cost per 100 ft	
			Mat'l, $	Labor, W-Hr	Mat'l, $	Labor, W-Hr
460	1-25 HP	Ea	1,100	25	350	25
	50 HP	Ea	1,400	40	520	25
	75 HP	Ea	2,400	50	700	25
	100 HP	Ea	2,400	55	1,200	45
	150 HP	Ea	4,200	80	1,400	50
	200 HP	Ea	4,500	85	2,300	60
	250 HP	Ea	6,400	105	2,900	75
	50 Amp Feeder (25 KVA)	Ea	460	25	350	25
	100 Amp Feeder (50 KVA)	Ea	680	30	520	25
	200 Amp Feeder (100 KVA)	Ea	940	45	1,200	45
	300 Amp Feeder (150 KVA)	Ea	2,200	55	1,400	60
	Welding Receptacle	Ea	900	30	350	25
4000	350 HP	Ea	10,800	50	900	110
	1250 HP	Ea	11,000	60	1,500	120
	2500 HP	Ea	13,000	75	3,500	140

Description	Unit	Mat'l, $	Labor, W-Hr
Active areas lighting & 110V outlets including feeder, transformer, panel, lights and receptacles - process, utilities and tank farm pumps areas	$100ft^2$	300	15
Inactive areas lighting, including feeder transformer, panel & lights - general yard & tank farms	$100ft^2$	100	5
Misc. 110V identifiable loads, valve operators, solenoids, etc., including feeder, transformer, panel and wiring devices	Ea	150	10
Simple interlocks, including one control device, pressure switch or equivalent, one interlock wiring, conduit, etc.	Ea	1,200	80
Complex interlocks required for large complex equipment such as large compressors, centrifuges, kilns, etc.	Ea	3,500	200
Ground connections including prorata of ground loop	Ea	100	6
13.8KV/480V 1000 KVA substation including 200ft of 13.8KV feeder and 4-200ft 480V feeders, complete with high and low voltage protective switch gear - without building	Ea	200,000	1,200
Same as above except for 2000KVA	Ea	250,000	1,500
13.8KV/4.16KV 3000KVA substation including 200ft of 13.8KV feeder and high and low voltage protecting fused switches - without building	Ea	185,000	1,200

Frequently the distribution system is at 13.87 kv and large motors are run at 4.1 kv. In those cases, the cost of a 13.8- to 4.1-kv substation must be included in the estimate.

Definitive Estimate

Definitive and/or appropriation type estimates can be developed in four or six hours with the information normally provided in a Phase I package, i.e., P&ID's, equipment list, plot plans, arrangement drawings and single line electrical diagram.

- Count all electric loads, 460 v and above, as before and classify by size.
- Make allowance for electric motors not included in P&ID's such as fans, HVAC equipment, motor operated valves and doors using 460 v, etc.
- Determine distance from each motor to motor control center, follow pipe racks routing and allow at least 10% for miscellaneous turn and deviations from the shortest route. The maximum distance should not exceed 500 ft, otherwise several transformers and/or motor control centers at separate locations may be required.
- Using an average length for the plant or for each area is a valid way to reduce estimating time.

NOTE: Motors over 460 v should always be considered individually.
- Calculate the total load as follows:

1 hp	1 kva
Active areas lighting	5.0 Watt/ft^2
Inactive areas lighting	1.0 Watt/ft^2
Electric heat tracing	15.0 Watt/ft
Building lights and miscellaneous	50 kva per building
Add 50% for future and miscellaneous	

NOTE: The minimum 480 v transformer size should be 1,000 kva. Reliability considerations may dictate the use of two 1,000-kva units. The maximum 480 v transformer size should be 2,000 kva.

- Assume one welding receptacle for every 5,000 ft^2 of process area with a minimum of one for each operating floor and utility area.
- Determine area requiring lighting - process and utilities.
- Determine number of interlocks for P&ID's and/or single line diagram.

- Consider one grounding connection for every equipment item and add 20% for grounding of miscellaneous structures.
- Apply the pertinent units from Tables 13.24.

13.7 Instrumentation Estimating Procedure

Introduction

An accurate cost estimate for the instrumentation account requires a substantial engineering effort:

- Reviewed P&ID's.
- Detailed take-offs.
- Current material prices.

It is a time-consuming effort and the cost can be justified only for an approved project.

At the other extreme, the instrumentation account could be estimated as a percentage of the equipment cost. However, this method is very inaccurate; published information shows a range of 15% to 50% of the equipment cost.

The cost of instrumentation is related mainly to equipment type and count rather than to equipment cost.

- Except for the size of the associated control valves, the instrumentation of a small piece of equipment is essentially the same on a large one.
- The unit cost of field mounted instruments varies with the materials of constructions; however, the cost of panel mounted instruments, DCS's, PLC's, etc., does not.
- The cost of handling, calibrating, installing and loop checking depends only on the number of instrument items and is in no way related to the cost of the equipment.

This procedure presents a quick and reasonably accurate method to estimate the cost of the instrumentation account at various levels of engineering development with sufficient details for subsequent cost tracking and project control.

Take-Offs

When reviewed P&ID's and time are available, count:

- DCS points - diamonds and squares.
- DCS wiring connections - some diamonds and squares may have more than one connection.
- Field instruments - balloons only (don't include diamonds and squares).
- Air hookups - 0.0 per self-contained CV; 1.0 Per XVC; 1.5 Per CV.

See Fig. 13.11 for a typical P&ID representation and take-off guideline.

NOTE: Different instrument engineers follow different systems to identify field instruments affecting the balloon count. A discussion should be held with the acting instrument engineer to compare the P&ID's with Fig. 13.11 and try to reconcile the take-offs.

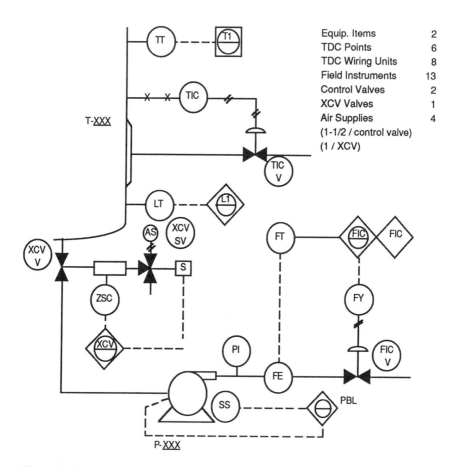

Equip. Items	2
TDC Points	6
TDC Wiring Units	8
Field Instruments	13
Control Valves	2
XCV Valves	1
Air Supplies	4
(1-1/2 / control valve)	
(1 / XCV)	

Figure 13.11 Instrumentation estimating procedure take off guideline.

Semi-Detailed Estimating System

When reviewed P&ID's and/or time are not available, reasonably accurate counts can be developed from the equipment list using the following densities, which are representative of organic chemical plants and include both batch and continuous operations.

		No./Equip Item.
- DCS Points	Batch Process	3.6
	Total Plant	2.0
	Continuous Process	1.7
	Storage Areas	1.0
- DCS Wiring Units	Batch Process	5.4
	Total Plant	3.0
	Continuous Process	2.5
	Storage Areas	1.5
- Field Instruments	Batch Process	12.0
	Total Plant	7.5
	Continuous Process	7.0
	Storage Areas	6.5
- Air Hookups	Batch Process	1.8
	Total Plant	1.0
	Continuous Process	0.9
	Storage Areas	0.5

Pricing

After the take-offs are completed, the material and labor costs are developed with the following units:

Materials

- DCS Hardware Including Computer		$1,800 per point
- Field Instrumentation	CS equipment	$650 per balloon
	low alloy equipment	$850 per balloon
	high alloy equipment	$1000 per balloon

Installation Subcontract

- Electrical Installation
    1. Includes all installation, materials, | 50% of field instru-
       conduits, wiring, multiconductor cables | mentation cost or
       junction boxes, termination panel, | $325 and 18 hr per
       miscellaneous supports, etc. | wiring unit.
    2. Trays for multiconductor cables not included. |
    3. Includes DCS installation. |
    4. Material/labor split approximately 20/80. |
    5. Labor on subcontract basis at $42/hr. |

- Pneumatic Installation
    1. Includes piping, tubing, valves, | 30% of field instru-
       miscellaneous supports, etc, from | mentation cost or
       instrument air header to instrument. | $220 and 24 hr per
    2. Air header not included. | air hookup.
    3. Material/labor split approximately 20/80. |
    4. Labor on subcontract basis at $42/hr. |

- Mechanical Installation and Miscellaneous
    1. Receiving and storage. | 30% of field instru-
    2. Calibration. | mentation cost.
    3. On-line and off-line installation, |
       including miscellaneous materials. |
    4. Checkout. |
    5. Material/labor split approximately 20/80. |
    6. Labor on subcontract basis at $42 hr. |

When the only information available is the equipment list, a good rule of thumb is to use $17,000 per equipment item. This number is based on an average case of:

    2.0 DCS points. | per equipment item.
    3.0 DCS connections. |
    7.5 Balloons. |
    1.0 Air hookup. |

This number is very consistent with the actual cost of several organic chemical plants and several definitive estimates.

It must be noted that in the case of a highly automated 100% batch process plant, the cost could be as high as $25,000-30,000.

13.8   Engineering Hours Estimating System

Introduction

Estimating engineering costs as a percentage of the direct project costs is a very common practice in conceptual and preliminary work.   It is very simple but also very risky.   The engineering hours related to any piece of equipment have very little relation to its cost.

For instance, the engineering hours related to a small or medium sized carbon steel reactor are essentially equal to those related to a larger Hastelloy C reactor.   However, the difference in cost can be a full order of magnitude. Engineering hours relate best to the type of equipment and conditions specific to each project.

The estimating system presented here proposes a relatively simple but accurate method for estimating engineering hours based on the equipment list and a general knowledge of the project particulars.   It was developed from published data and a thorough analysis of nine actual projects.   The analysis of the data resulted in several adjustment and correction factors (some empirical, some theoretical) that recognize, and try to compensate for, the effect of conditions specific to each project, such as:

- New technology/repeat plant.
- New site/retrofit.
- Phase I by Owner/by others.
- Plant size and complexity.
- Engineering contractor size.
- Non-process buildings.
- Execution approach.

The system includes procedures for estimating:

- Hours at engineering contractor's office.
- Hours to prepare a Phase I package.
- Hours required by the Owner to monitor contractor's work.
- Hours required to do the engineering with in-house resources.

In addition to the nine original study cases, the system has been repeatedly tested, sometimes against final project returns, other times against contractors' estimates. The results, tabulated in Table 13.25, are very good. When tested against final returns, the estimates prepared with the system have invariably been more accurate than the contractors' estimates prepared at the same stage of the project.

Table 13.25 Engineering Hours Estimating System Perfomance

Type of plant	Contractor Size	Actual hrs Phase II	Equip Count	Hr per Item	Accr'y Estm/Actual
Retrofit	Small	32,570	66	493	0.99
Grass roots	Large	416,600	540	771	1.01
New unit	Small	12,870	21	613	1.02
Grass roots	Small	78,200	118	663	1.02
New unit	Small	76,050	165	461	1.00
Grass roots	Large	262,600	307	855	0.96
Retrofit	Small	27,210	74	368	0.99
Retrofit	Large	172,760	123	1,404	1.03
New unit	Large	10,930	6	1,822	0.86
New unit	Small	19,500	45	433	1.01
Retrofit	Small	11,000	33	333	1.24
New unit	Large	67,300	53	1,270	1.08
New unit	Small	9,520	9	1,058	1.22
Retrofit	Large	18,540	5	3,708	0.92
Grass roots	Large	162,300	273	595	1.05
Grass roots	Large	34,270	52	659	0.94
Estimate	Small	15,950	45	355	1.22
Retrofit	Small	87,470	201	435	1.09
Retrofit	Small	18,600	121	154	1.08
Grass roots	Large	107,200	134	800	1.03
New unit	Small	7,200	21	342	1.00
New unit	Small	22,000	100	220	1.14
Grass roots	Large	159,800	223	717	0.94
Retrofit	Large	22,000	95	232	1.05

NOTE OF CAUTION: All the original study cases predate the widespread use of computer-aided design programs. Although the author expected to see a dramatic reduction in engineering hours, that did not materialize for several years. However, in the last year or so, as engineers and designers become proficient with the system, the 3D CAD in particular, some contractors appear to be spending as much as 20-25% fewer hours. It has become very important in the contractor selection process to pay special attention to this area of contractor expertise.

Hours at Engineering Contractor's Office

Summary

The objective of this procedure is to develop a quick and reliable estimate of total contractor home office hours with minimum information.

Information Required

- Equipment list including installed spares:
    1. Preliminary size.
    2. Pressures.
    3. Preliminary horsepower.
- Special conditions:
    1. Continuous/batch process.
    2. Type of instrumentation.
    3. Distribution/collecting systems.
    4. Piling/site preparation.
    5. Non-process buildings.
    6. Retrofit.
- Preliminary Execution Plan:
    1. Phase 0/Phase I by others.
    2. Size of contractor.
    3. Direct hire/subcontract.

Procedure

- Computation of base hours - Tables 13.26, 13.27 and 13.28.
- Adjustment of base hours - Tables 13.29 and 13.31.
- Corrections for equipment size, contractor size and number of equipment items - Tables 13.30 and 13.31.

Basis of Estimate

Table 13.26 - Engineering Contractor Hours - Equipment

This table shows the home office hours required by a large contractor to design and procure the different types of equipment normally encountered in a chemical plant. The hours comprise all technical and clerical activities from the start of process design to project closeout. They are intended for a continuous operation, single unit plant with electronic DCS instrumentation built on a direct-hire basis on a clear site and include process-related buildings, such as a motor control center and control room.

The units include hours for activities that are frequently performed either by subcontractor or directly by the Owner. These activities fall into the following areas:

- Process design.
- Estimating, schedule and cost control.
- Procurement.

It must also be noted that the base hours include secretarial and clerical work. The author's suggested practice is to consider these hours with the contractor overhead rather than a reimbursable cost. Table 13.29 includes the recommended adjustment.

In order to simplify the work, the equipment has been classified into nine groups; the hours for each group represent the average of the various items included.

Table 13.27 - Engineering Contractor Hours - Miscellaneous

This table was developed to account for the home office work associated with activities that are required on most projects but are not directly related to equipment.

Table 13.28 - Engineering Contractor Hours - Normal Breakdown

This table shows the normal breakdown of the home office hours by discipline for both liquids and solids handling process.

Table 13.29 - Engineering Contractor Home Office Adjustments

This table accounts for the deviations from the typical case encountered in practically all cases.

Table 13.30 - Engineering Contractor Hours Overall Correction Factors

This table includes the across-the-board corrections that must be applied to the total hours to account for:

- Complexity as a function of the number of equipment items, including installed spares.
- Size of equipment as a function of the average horsepower of process equipment motors (pumps, agitators, compressors).
- Size and sophistication of the engineering contractor.

Table 13.31 - Engineering Contractor Work Hours Breakdown, Adjustment and Correction Worksheet

This form is suggested to facilitate and document the adjustments and corrections required for each project.

Computation Procedure

Step 1 - Compute Base Hours

Base hours are all the home office hours required to perform all the engineering, procurement, subcontracting, project management, estimating and cost control for a normal case plant. They are computed as follows with the aid of Tables 13.26, 13.27, 13.28 and 13.31.

1. Count and classify the equipment under the categories listed in Table 13.26. Installed spared items must be included in count.

    In retrofit projects, it is very convenient to compute the base hours by groups of equipment that reflect the scope of the retrofit work, such as:

    - New equipment in existing structures.
    - New equipment in new structures.
    - Relocated existing equipment.
    - Existing equipment in-place.

2. Apply the corresponding unit hours and compute the base hours for equipment.

Table 13.26 Engineering Contractor Hours – Equipment

Ref. Equipment Description	Proc. Eng	Non Process			Total
		Proj Mgmt & Control	Design	Purch & Exp	
1 - Process furnaces	750	670	2,770	440	4,630
- Reactors 2 - Tray towers - Direct fired heaters - Large process incinerators	390	440	1,710	240	2,780
- Packed towers - Custom designed press. vessels > 4'D - Reboilers - Package boilers 3 - Motor driven process compressors - Turbine driven process compressors - Complex multi unit packages - TDC/PLC and batch controllers - Rotary driers/filters	270	190	960	140	1,560
- Custom designed press vessels < 4'D - Atm stg tanks & proc. vessels - Instrument air package - Emergency generators - Refrigeration packages - Multi stage ejector systems 4 - Vacuum pumps - Environmental scrubbers - Belt & screw conveyors - Mills - Thickeners - Vibrating screens	160	110	550	80	900
- Shell & tube heat exchangers - Eductors - Single stage ejectors - Wiped film evaporators 5 - Cooling towers - Drum driers - Elevators - Crushers - Blenders - Kneaders	120	70	350	60	600

Table 13.26   (Continued)

| Ref. | Equipment Description | Proc. Eng | Non Process | | | Total |
			Proj mgmt & Control	Design	Purch & Exp	
	- Turbine driven pumps & blowers					
	- Mixers and agitators					
	- Simple incinerators					
	- Flare stacks					
6	- Double pipe heat exchangers	60	40	240	40	380
	- Centrifuges					
	- Dust collectors					
	- Packaging machines					
	- Mechanical feeders & flakers					
	- Hoppers					
7	- Motor driven pumps & blowers	20	50	140	10	220
	- Fans & extractors					
	- Hoist and trolleys					
	- Rotary valves					
	- In-line filters					
	- Small separators, K.O. pots					
8	- Off the shelf press. vessels	10	10	30	10	60
	- Heating coils/plate & bayonet heaters					
	- Sample coolers					
	- Electric heaters					
	- Static mixers					
	- Mobile equipment					
	- Portable air conditioners					
9	- Misc. non equip. items, tote bins/buggies/scales/vent filters	5	5	10	5	25

Table 13.27 Engineering Contractor Hours – Miscellaneous

		Non Process			
Ref. Equipment Description	Proc. Eng	Proj mgmt & Control	Design	Purch & Exp	Total
Subcontract award and administration					
101 Lump sum	–	120 to 240	150 to 300	150 to 300	420 to 840
102 Reimbursable	–	240 to 480	100 to 200	200 to 400	540 to 1080
Systems shared by multiple unit plants					
103 Steam distribution & condensate system	80 to 160	100 to 160	400 to 600	80 to 120	660 to 1040
104 - Vent collection system	60 to 80	50 to 100	240 to 360	60 to 80	410 to 620
105 - Cooling water distribution system	40 to 80	50 to 80	200 to 300	40 to 60	330 to 520
106 - Waste water collection system	–	40 to 60	200 to 300	40 to 60	280 to 420
107 - Fire protection system	–	20 to 40	120 to 200	20 to 40	160 to 280
108 - High/medium voltage elect. distribution	–	40 to 60	200 to 300	40 to 60	280 to 420
109 - Interconnecting pipes & pipe rack	40 to 80	120 to 200	800 to 1200	80 to 120	1040 to 1600

Table 13.27 (Continued)

Ref.	Equipment Description	Proc. Eng	Non Process			Total
			Proj mgmt & Control	Design	Purch & Exp	
	Main Substation					
110	Estimated cost not available	–	180	1,000	120	1,300
111	Basic eng. only - detailed eng by subcontractor - 5% of estimated cost broken down as follows	–	0.25	0.60	0.15	1.00
102	All eng by contractor - 15% of estimated cost broken down as follows	–	0.2	0.7	0.1	1.00
	Miscellaneous Site Work					
113	Piling, earth movement access roads, utility, supplies, etc. - 15% of estimated cost broken down as follows	–	0.2	0.75	0.05	1.00
	Non-Process Buildings					
114	Basic eng. only, details by design/build S.C.- 5% of estimated cost broken down as follows	–	0.25	0.60	0.15	1.00
115	All eng by contractor - 10% of estimated cost broken down as follows	–	0.20	0.70	0.10	1.00

3. Select the applicable units from Table 13.27, apply the corresponding unit hours and add to the equipment hours for the total base hours.
4. Use Table 13.28 to break down the total hours by discipline and enter in the first column of Table 13.31.

IMPORTANT NOTE: Occasionally a project involves repetitive steps with identical equipment, instruments and piping arrangements. Most of the detailed engineering developed for the first unit will be used for all such units with substantial time savings. So the estimator, after estimating the first units, must use judgment to make the proper adjustments to the other units.

Step 2 - Compute Adjusted Hours

Adjusted hours are the home office hours that would be required by a large engineering contractor to perform a specific scope of work for a plant with normal sized equipment. They are derived from the base hours, adjusted to take into consideration the specific situation; i.e., extent of process design required, type of process and instrumentation, special site conditions, extra retrofit work, existing facilities, Owner's participation, etc.

Table 13.28 Engineering Contractor Hours - Normal Breakdown

Activity	Fraction of Subtotal		"Normal" Percent of Total
Subtotal Process Engineering (from Table 13.26)			11-13%
Project Management and Control			
Project Management/Engineering	0.45		
Estimating	0.15		
Cost Control	0.25		
Scheduling	0.15		
Subtotal Project Management and Control (from Table 13.26)			15-18%
Detailed Engineering	Fluid[1]	Solid[2]	
Civil/Structural/Architectural	0.15	0.20	
Electrical	0.13	0.16	
Instrumentation	0.13	0.12	
Piping [3]	0.35	0.18	
Equipment/HVAC/Mechanical	0.10	0.20	
Engineering Clerical	0.14	0.14	
Subtotal Detailed Engineering (from Table 13.26)			58-64%
Procurement			
Purchasing	0.65		
Expediting	0.35		
Subtotal Procurement (from Table 13.26)			8-11%
GRAND TOTAL			100%

[1] For fluid handling plants. Based on average of actual cases
[2] For solid handling plants - best estimate
[3] Piping work-hours breakdown

P&ID drafting	0.07
Layouts	0.05
Model	0.10
Orthographics	0.33
Material take-offs	0.10
Stress analysis	0.05
Isometrics	0.30

Once the basic hours have been computed and broken down by discipline, the adjusted hours can be computed as follows with the aid of Tables 13.29 and 13.31:

1. From Table 13.29, Items 1 to 15 only, identify the conditions applicable to the project, show the adjustments in the designated column of Fig. 13.31 and compute the adjusted hours for process, design and support, and procurement.
2. Make intermediate adjustment of project and control hours as directed in Item 16, Table 13.29.
3. From Table 13.29, Items 17 to 20, identify the other conditions applicable to the project and compute the adjusted project and control hours.
4. Compute the total adjusted hours.

NOTE: This step requires engineering judgment and a good understanding of the site conditions and the way in which the project will be executed.

Step 3 - Compute Corrected Hours

Corrected hours are the home office hours that would be required by a given engineering contractor to perform the specific scope of work for a particular plant. They are derived from the adjusted hours, corrected as follows by the factors included in Fig. 13.30, which are based on the equipment count, equipment size and contractor size.

1. Determine the total number of pieces of equipment including installed spares.

NOTE: Non-equipment items in Group 9, Table 13.26 must not be included in the count.

2. From Table 13.30, determine the equipment count factor (F_e).
3. Count electric motors and compute total horsepower.

NOTE: Count must be limited to major equipment: i.e., pumps, agitators and compressors. Rotary valves, chemical feed pumps and the like must not be included.

4. Compute the average horsepower (total hp/number of motors) and determine the correction factor (F_s) from Table 13.30.
5. From Table 13.30, determine the contractor size factor (F_c).
6. Combine the three correction factors into a single factor and use the last column of Table 13.31.

Table 13.29 Engineering Contractor Home Office Adjustments

Items	Adjustments	Comments
1 Phase I design by others	Multiply Process hr x 0.3 Instrumentation hr x 0.9 P&ID drafting hr x 0.4 Piping layout hr x 0.6 Equip./HVAC/Mech hr x 0.9	
2 Phase 0 design by others	Multiply Process hr x 0.70	
3 Batch process	Multiply Process hr x 0.7 Instrumentation hr x 1.2	Apply only to process units affected
4 No model required	Multiply Piping model hr x 0.0	
5 Isometrics by subcontractor	Multiply Piping isos hr x 0.0-0.25	Some area isometrics and/or undimensioned isos may be required
6 Heavy retrofit Removal & installation of equipment in operating areas engineered while existing equipment is in operation & requiring plant shutdown	Multiply C/A/S hr As required Elec hr As required Piping orthos hr x 1.8-2.5	Corrections to be applied only to areas requiring the retrofit

Table 13.29 (Continued)

Items	Adjustments	Comments
7 Light retrofit	Multiply	
Installation of equipment in areas that may be cleared before design	Piping orthos hr x 1.3-1.5	Same as above
8 Relocate existing equipment	Multiply	Apply correction only to particular equipment.
	Mech/HVAC/Eq. hr x 0.25	
	Procurement hr x 0.2	
	Adjusted process hr x 0.6	
9 Existing equipment in place	Multiply	Apply correction only to particular equipment. If existing piping, instruments or electrical are to be reused, adjust accordingly.
	Adjusted process hr x 0.6	
	C/A/S hr x 0-0.25	
	Mech/HVAC/Eq. hr x 0.25	
	Procurement hr x 0.2	
	Piping hr As required	
	Electr. hr As required	
10 Existing buildings and/or structures	Multiply	Apply only to equipment affected. Check for retrofit & adjust if applicable.
	C/A/S hr x 0-0.25	
11 Old plant site	Multiply	To cover for potential under-ground problems
	C/A/S hr x 1.1	
12 Existing utilities	Multiply	
	Piping orthos hr x 0.8	
13 Non-reimbursable clerical	Multiply	
	Prorated clerical & admin hr x 0.2	Prorate as follows: $\dfrac{\text{Adjusted design hr}}{0.86} \text{ x } 0.14$

Table 13.29 (Continued)

Items	Adjustments	Comments
14 Procurement by owner and/or subcontractor	Multiply Adjusted procurement hr x 0.2-0.5 Piping MTO hr x 0.2-0.5	
15 Instrument engineering by owner	Take credit based on Owner's involvment	
16 Proj. Mgt & controls Proj/Est/Cost Control/ Scheduling	Once the process engineering, detailed engineering & procurement hours have been adjusted, the Project Management and Control hours are adjusted to maintain the same proportion calculated in the base hours total. The adjusted hours are then broken down in the fractions indicated in Table13.28 and each fraction adjusted as follows:	
17 Initial cost estimate by Owner	Multiply Estimating hr x 0.5	
18 Cost control by Owner and/or subcontractor	Multiply Cost Control hr x 0.2 – 0.5	
19 Scheduling by Owner and/or subcontractor	Multiply Schd. hr x 0.2 – 0.5	
20 Project Mgt/Eng for small projects	Set minimum hr for project mgt/eng based on project duration & Owner participation in detailed engineering	Below certain project size a minimum PM/PE attention is required based on project duration

Table 13.30 Engineering Contractor Hours Overall
Correction Factors

Complexity Factor

Equipment Count	Multiplier (F_e)
10	1.80
25	1.50
50	1.30
100	1.05
125-175	1.00
200	1.03
250	1.04
300	1.08
600	1.30

Equipment Size Factor

Ave. Motor hp Process Equipment	Multiplier (F_s)
10	0.90
10–15	0.95
15–50	1.00
50–100	1.05
100	1.10

Contractor Size Factor

Large Contractor	Multiplier (F_c)
Over 300 people, inflexible organization, sophisticated control systems	1.00

Small Contractor	
Flexible, simple control, versatile Project Managers	0.80-0.90

Table 13.31 Engineering Contractor Work-Hours Breakdown, Adjustment and Correction Worksheet

Job: _____

	Base hours		Adjustments	Adjusted hrs	Corrected
Process					
Project & Control Proj. Mgt./Eng			From item 16, Table 13.29		
Estimating					
Cost control					
Scheduling					
Sub total					
Design & Support Civil, arch. & struct.					
Electrical					
Instrumentation					
Piping					
Equip./ HVAC/ Mech.					
Clerical & admin.					
Sub total					
Procurement Purchasing					
Expediting & insp.					
Sub total					
Total					
Correction Factor			±	Equip. cnt. (Fe) __ Equip. size (Fs) __ Contr. size (Fc) __	

IMPORTANT NOTE: Experience shows that when a contractor is retained to perform detailed engineering only on specific disciplines, with the Owner retaining the overall responsibility, there is very little sensitivity to the equipment count.

Hours to Prepare Phase I Package

The hours required for the preparation of a Phase I package are derived from the base man-hours (before adjustments or corrections) in the contractor's hour estimating procedure as follows:

Normal Project

First design of a known process. No previous Phase I available.

Discipline	Fraction of Contractor Base Hours
Process	0.60
Instrumentation	0.10
Piping	0.05
Equipment/HVAC/mechanical	0.05
Project and Miscellaneous	Add 20%

Repeat Project

Phase I design(s) available from previous plant(s).

Discipline	Correction Factor to Normal Project
Process	0.4-0.8
Instrumentation	0.4-0.8
Piping	0.8-1.0
Equipment/HVAC/mechanical	0.6-0.8
Project and Miscellaneous	Add 20%

New Process

Process design developed from pilot and R&D data.

Discipline	Correction Factor to Normal Project
Process	1.5-2.0
Instrumentation	1.2-1.5
Piping	1.0-1.2
Equipment/HVAC/mechanical	1.0-1.2
Project and Miscellaneous	Add 20%

Preparation of a Phase 0 package requires approximately 40% of Phase I. This must be taken into consideration when the Phased I work is based on an existing Phase 0.

Hours to Monitor Contractor's Work

The contractor doing the engineering work requires a certain degree of monitoring and supervision and, in some cases, even direction from the Owner. This is especially true when the contractor performs its work under a cost reimbursable contract.

This procedure proposes a rational method that relates the required Owner's hours to the estimated contractor's man-hours in a consistent and rational manner.

Discipline	Fraction of total contractor hours for each discipline
Process engineering	0.20
Project management/engineering	1.00
Estimate/schedule/control	0.20
Instrumentation	0.20
Electric/mechanical/procurement	0.05
Overhead administration	*
Auditing	*

*Estimate each project based on desired participation.

Hours for In-House Engineering

For a small project of the right complexity, engineering can be most effectively done at the plant level, provided the proper resources are available. The hours required can be derived from the estimating system for contractors' home office hours as follows:

Base Hours

- **Procurement and Clerical** - Unless additional help is specifically required, this work will normally be absorbed by the regular plant staff.
- **Estimating, Cost Control and Scheduling** - The work in these areas will be the bare minimum and the project engineer would perform it.
- **Adjustments to the Design** - As required in Table 13.29.

Correction Factors

- **Equipment Count Factor** - Experience has shown that in-house engineering is less sensitive to low equipment counts. The following correction factors are suggested:

Equipment Count	Correction Factor
10	1.08
25-50	1.00
75	1.05

- **Equipment Size Factor** - Same as for contractor's man-hours.
- **Contractor Size Factor** - Because of the smaller size of the groups, the less formal communication and documentation procedures and the simple supervision structure, a plant engineering group is able to perform engineering even more effectively than a small contractor.
- A multiplier of 0.6 is warranted.

13.9 Field Costs

Labor Costs

The labor costs should be estimated in a manner consistent with the planned execution approach.

- If construction is to be executed on a direct-hire basis, the labor costs should include only the basic labor rate. All other labor-related costs (PAC's, fringe benefits, supervision, rental equipment, etc.) should be included with the field indirects.
- When construction is to be subcontracted, some accounts (civil, insulation and paint) are usually estimated on a total cost basis, combining both materials and labor. Other accounts (piping, electrical, instrumentation, equipment erection) are estimated and shown separately as material and labor. In that case, the labor should be loaded with the subcontractor's field indirects plus overhead and profit. The field indirects would reflect only the cost of construction management.

The Project Manager and/or Cost Engineer reviewing a cost estimate must establish whether the costs in the labor column reflect base rates, loaded costs or some intermediate value, such as base rate plus fringe benefits. Failing to do so would result in misleading material to labor ratios and incorrect cost analysis.

Table 13.32 indicates the labor rates and fringe benefits that should be used to prepare direct-hire estimates at several U.S. locations. These rates are based on current union rates (mid-1993) for pipefitters and represent a conservative average of the various crafts involved in construction.

The following is a typical subcontractor loading:

	Fraction
Base rate	1.00
Fringe benefits*	0.25
Supervision	0.15
Insurance and tax (PACs)**	0.25
Small tools and consumables**	0.10
Subtotal	**1.75**
Overhead and profit, 20%**	
TOTAL	**2.10**

* Use actual value when available.

** These are typical values. Actual values will change with site, contractor and business atmosphere.

When time permits, the loaded rate of local subcontractors should be determined prior to estimating.

Subcontractors' loading does not include the cost of the rental construction equipment such as cranes, welding machines, scaffolding and expensive tools exceeding a predetermined value. These costs are usually reimbursable based on the rental rate submitted by the subcontractor in the proposal.

A simple way to account for the cost of rental equipment used by the subcontractors is to add a discrete amount to the loaded cost. The following amounts are suggested:

- Equipment erection labor $8/hr
- Structural steel erection labor $5/hr
- Piping labor $3/hr
- Electrical/instrumentation labor $2/hr

Table 13.32 Typical Union Labor Rates, Pipefitters July 1993

Location	Base Rate, $/Hr	Fringe, $/Hr	Benefits, frac. of base
Atlanta, GA	18.45	4.06	0.22
Baltimore, MD	20.17	6.31	0.31
Billings, MT	19.00	4.25	0.22
Charleston, WV	19.63	3.75	0.19
Cheyenne, WY	15.82	3.98	0.25
Chicago, IL	24.55	5.57	0.23
Columbus, OH	21.07	6.29	0.29
Denver, CO	18.92	3.83	0.20
Detroit, MI	22.76	7.95	0.35
Houston, TX	17.51	4.05	0.23
Little Rock, AR	14.35	2.43	0.17
Milwaukee, WI	21.46	5.02	0.23
New Orleans, LA	16.80	3.10	0.18
Omaha, NE	19.39	4.18	0.22
Philadelphia, PA	25.27	9.60	0.38
Richmond, VA	15.40	2.30	0.15
Topeka, KS	16.13	2.04	0.13
Wilmington, DE	21.21	4.89	0.23
Nation Highest			
San Francisco, CA	35.55	17.63	0.50
New York, NY	27.30	18.31	0.67
Oakland, CA	31.07	8.77	0.20
Nation Lowest			
Columbia, SC	12.42	1.06	0.08
Fargo, ND	12.37	1.97	0.16
Jackson, MS	12.50	2.05	0.16
Corpus Christi, TX	13.26	1.53	0.12

The rates for non-union labor may be somewhat lower (5-10%) than for union labor but, generally, they are competitive and no credit should be taken in the estimate without an area labor survey.

The potential advantage of using non-union labor is in the lower percentage of time lost because of stringent union work rules. However, this advantage can easily be offset by lower productivity resulting from poor training of the non-union labor pool.

Field Indirects

As discussed in the preceeding paragraphs labor costs are defined either as the wages paid directly to the workers in a direct hire situation or the monies paid to the subcontractors to cover all their field expenses (except materials) plus overhead and profit. The field indirects are then, all other project-related costs incurred in the field in addition to the labor costs. The scope of the field indirects account in the estimate is naturally in inverse proportion to the scope of the labor costs.

Whether some of them are folded into the labor costs or not, the field indirects include cost items such as:

- Labor fringe benefits.
- Payroll added costs (PAC's).
- Supervision.
- Field offices, storage and other temporary facilities.
- Rental construction equipment.
- Miscellaneous tools and supplies.
- Lost time.
- Others.

The Field Indirects Checklist and Field Indirects Criteria in Appendix I are intended mainly as control tools to check contractors' detailed estimates and planning as well as to monitor their performance in reimbursable contracts. They can also be used as an estimating tool to develop in-house detailed or semi-detailed estimates. However, for in-house and preliminary estimates, sufficient accuracy can be achieved by rationing the field indirects to direct labor and subcontracts.

In lump sum contracts and subcontracts, even if all field indirects should be included, the Project Manager must ascertain that they really are and that adequate supervision and support are provided to insure proper execution. Appendix I is also a good tool for that purpose.

When construction is totally subcontracted the field indirects are a mixed bag; some are included in the subcontracts while others must be provided by the construction manager or the general contractor. The field indirects likely to be provided by the construction manager or the general contractor are identified in the Field Indirects Checklist.

The field indirects account for a detailed estimate, specifically in a large project, should be broken down into all its major components to insure that all costs are considered. In semi-detailed and preliminary estimates a breakdown is not required. Sufficient accuracy can be achieved by using the following factors:

- 95 - 120% of direct labor

    plus
- 15 - 20% of subcontract labor (loaded)

    plus
- 10 - 12% of subcontracts (material and labor)

The factors are inversely proportional to the construction hours.

Labor Productivity

General

Labor productivity is always a major concern to the estimator. Variations of up to 100% between two specific sets of circumstances are not unusual. With loaded labor costs running around 40% of total project costs, a 25% variation in productivity could impact total cost by 10%.

The cost impact of the productivity should be shown in the estimates as a line item with the field indirects account. This approach will permit a more consistent analysis of the materials to labor ratios and provide better project control by specifically identifying a potentially important source of abnormal project costs.

Factors Affecting Productivity

The Construction Industry Cost Effectiveness Study (CICE) sponsored by the Business Roundtable in the early 1980's indicated that, to a large extent, poor labor productivity is the result of inadequate management practices. Conversely, the risk of poor productivity can be minimized by effective project management. The purpose of this section is to give project managers a better understanding of the factors affecting productivity and provide guidelines to create, early in the project, an environment conducive to good labor productivity.

Table 13.33 lists the most important factors that affect productivity. Most of them are under control of owners and contractors and can be biased in the right direction by good project management.   The Owner can control or at least influence the first eight items by:

-   Awarding all construction work on competitive lump sum bases.
-   Maximizing the use of design-build contracts.
-   Breaking down work, as much as is practical, into discrete portions in order to maximize the use of local contractors using regular employees.
-   Setting a realistic schedule to avoid regular overtime and/or excessive staffing.
-   Demanding the highest safety standards from all contractors.
-   Promoting workers' motivation programs.
-   Implementing thorough procedures for screening and qualification of bidders.
-   Encouraging the merit shop and site agreement approach to construction.

The Owner has little or no control over the quality of the contractor's labor, supervision and organizational capabilities.   However, when contractors are selected through a careful and deliberate selection process, the most qualified are likely to be selected.

Neither the Owner nor the contractor can control factors like project size, type and location; nor can they control local weather or economic environment. However, they can minimize their adverse effects with good project planning and management and, in all cases, by recognizing the dangers and making proper productivity allowances in the estimate.

Basis of Construction Hours

Labor productivity should be measured against consistent standards. Everybody talks about Gulf Coast productivity and of using Gulf Coast unit hours in their estimates. Yet everybody seems to have a different set of Gulf Coast unit hours. In real life, many contractors have their own units and modify them periodically to reflect actual experience.

The unit hours used in all the estimating procedures are based on hours used by large contractors to estimate construction work on a direct-hire basis, reduced by 15% to allow for work done on competitive lump sum basis.

The unit hours shown in the concrete and insulation estimating procedures are intended for planning and scheduling only, since the unit costs are based on all inclusive subcontract prices.

Table 13.33 Factors Affecting Productivity

		Poor Productivity	Good Productivity
1.	Contracting approach	Separate engineering & construction contractors	Design-Build
		Negotiable	Competitive
		Reimbursable	Lump Sum
2.	Construction approach	Direct hire	Subcontract
3	Type of contractor	Large national, general contractor	Small local subcontractors
4.	Safety standards	Low	High
5.	Contract size	Large	Small
6.	Schedule	Fast track/O.T.	Normal
7.	Work rules	Union	Merit shop
8.	Labor source	Union hall/open market	Regular employees
9.	Labor quality	Poor	Good
10.	Suppervisor quality	Poor	Good
11.	Field organization	Poor	Good
12.	Site location	Remote	Accessible
13.	Site condition	Clustered	Clear
14.	Type of work	Retrofit	New unit
15.	Project size	Large	Small
16.	Weather	Poor	Good
17.	Local economy	Depressed	Booming
18.	General area productivity	Poor	Good

Productivity Adjustments

Conceptual and preliminary estimates normally would not contain sufficient detailed information for making any meaningful productivity correction to the labor accounts. Exposure to low productivity should be covered by the judicious evaluation of the contingency allowance.

Definitive and appropriation type estimates should consider project specific conditions with potential effects on productivity and incorporate provisions to minimize their adverse effects and/or cover their cost impact. As mentioned previously, most of the potential adverse effects can be eliminated or, at least, minimized by good project execution planning and management.

Some of the adverse conditions are inherent to the specific sites and/or project and cannot be eliminated through project planning and management. In these cases, the following multipliers should be applied to the estimated hours:

Condition		Multiplier	Comments
Retrofit and/or clustered Areas		1.20-1.40	To be applied only to work and areas specifically affected
Bad weather		1.05	Bad weather has more effect on schedule than on productivity. Apply factor to anticipated bad weather periods.
Second shift		1.10	To be applied only to work on second shift.
Unavoidable extended scheduled overtime:			
50 hr week	Up to 2 weeks	1.05	Even if the overtime is required
	2-4 weeks	1.10	in only one area and/or for one
	Over 4 weeks	1.20	particular craft, frequently it becomes necessary, to maintain labor peace, to place the entire project on overtime.
60 hr week	Up to 2 weeks	1.10	
	2-4 weeks	1.15	
	Over 4 weeks	1.30	

It must be noted that the cost of overtime extends far beyond the productivity loss since the rates of overtime work are at least 1.5 times the regular rate.

13.10   Adjustments

Resolution Allowance Criteria

When an equipment item is priced on in-house information (curves or past experience) and/or budget prices from vendors, the cost is developed without the benefit of a design specification and can either be high or low.

When the price is the result of a formal quote based on specifications developed early in the project, the detailed design, safety reviews and normal project development will usually result in extra costs:

- Additional recycle streams and instrumentation requires extra nozzles.
- Pumps may require different types of seals.
- Gasket and seal materials may need upgrading.

The same situation occurs when commodities (concrete, structural steel, electrical, etc.) are priced. If they are factored from the equipment account or derived from the actual cost of other projects, they could either be high or low. However, if they are developed thorough take-offs, from drawings at various stages of completion, the errors will be omissions rather than additions and some growth in quantities is expected. The amount of growth will be inversely proportional to the completion of engineering.

The expected increases in the equipment and commodity accounts should be considered a part of the base estimate rather than contingency. They should be included in the body of the estimate and identified as Resolution Allowances. The purpose of this guideline is to provide rational criteria to determine the appropriate resolution allowance on a consistent basis.

The recommended approach for semi-detailed estimates is to apply the following typical factors used by many contractors in the chemical industry.

Equipment

- 3-5% to equipment costs obtained through formal quotes only.
- 0% to equipment costs obtained from curves or budget quotes.

Commodities

- 20-30% on estimates based on take-offs from preliminary P&ID's and arrangement drawings (typical preliminary estimate).
- 10-20% on estimates based on take-offs from approved-for-design P&ID's and arrangement drawings (typical definitive estimate).
- 5-10% on estimates based on detailed drawings and bill of materials (typical engineering estimate).
- 0% on factored estimates.

Escalation

All estimates must be adjusted for the expected inflation to bring the cost as close as possible to the actual cost at the time of completion.

The simplest approach, quite adequate for small projects, is to apply the current inflation projection to the mid-life of the project; i.e., if the estimated duration is 14 months and the projected annual inflation rate is 6%, the escalation adjustment will be:

$$(14 \text{ months} \div 12 \text{ months/year}) \times 0.5 \times 6\% = 3.5\%$$

On large, extended projects, where the escalation allowance could become a substantial number, it is more appropriate to calculate the escalation for each of the major cost accounts based on the projected inflation rate and the project schedule. By doing so, every cost account will be escalated to the time in which most of the actual expenditure is projected.

Special care must be taken in escalating labor costs since the rates are determined by labor agreements lasting two or three years and it is quite possible that no change in rate occurs during the project or that a step change will occur before construction begins.

Contingency Determination

As in the case of resolution allowance, the required contingency is inversely proportional to the level of engineering completion. The contingency philosophy changes from company to company; it could even change from project to project. Some companies have an unrealistic approach and take an adamant attitude against overruns; they want 90/10 estimates. Other companies are more realistic, are willing to take reasonable risks and are satisfied with 50/50 estimates. The 90/10 estimates require substantially higher contingencies, as do estimates based on limited engineering.

The following levels of contingency recommended for various stages of engineering completion have been developed by the application of Hackney's definition rating method (Hackney, J. W. *Control and Management of Capital Projects*. 2nd edition, McGraw-Hill, Inc., 1992) to the scope of each engineering stage. They have been successfully tested against the recommendations developed through very sophisticated methods:

Engineering Level		Contingency Probability	
		50%	90%
Conceptual design	High	60%	137%
	Avg.	47%	107%
	Low	30%	60%
Phase 0 design	High	26%	58%
	Avg.	22%	50%
	Low	17%	39%
Phase I design	High	13%	29%
	Avg.	10%	22%
	Low	8%	17%
Basic engineering	High	10%	20%
	Avg.	8%	18%
	Low	6%	12%
Detailed engineering	High	6%	12%
	Avg.	4%	9%
	Low	3%	6%

The above contingencies are to be applied after the base estimate has been corrected with resolution allowances as previously indicated.

IT IS VERY IMPORTANT TO NOTE THAT THE BASE ESTIMATE IS NOT A FIXED NUMBER. IT MUST BE RE-EVALUATED PERIODICALLY TO REFLECT ACTUAL PROJECT PERFORMANCE AND ANY ADDITIONAL INFORMATION ACQUIRED DURING EXECUTION.

13.11   Quick Estimate Checks/Conceptual Estimating

General

The shortcut techniques included here are very useful for quickly checking estimates as well as developing conceptual estimates. Most have been derived from the analysis of several organic chemical plants, built or estimated; others come from personal observations. Table 13.34 reflects the analysis of nine actual chemical projects and five estimates of the following common characteristics:

- **Materials of Construction** - Mostly low and medium alloy steels.
- **Instrumentation** - Partially automated. DCS with approximately two points per equipment item. Seven or eight P&ID field balloons by equipment item.
- **Piping** - 2.5 to 3 in. average diameter.
- **Structural Steel** - Most process equipment on high bay structures.
- **Construction Labor Rate** - $40 per hr (loaded).
- **Contracted Engineering Rate** - $65 per hr (loaded).

The table is based on the following definitions:

Equipment Count

- All process - and utility-related items, including installed spares and in-line filters are included in the count.
- Items such as conservation vents, service hoists, buggies and moving equipment are not counted.
- Skid-mounted packages are considered as one item.
- Vendor designed packages broken down and delivered in pieces to be installed and piped in the field are counted based on the number of pieces.

Total Installed Cost (TIC) Scope

The following cost items are not included in the costs:

- Extraordinary site work such as extensive earth movement, piling, retention ponds, utility supplies, access roads, etc.
- Buildings other than sheltering structures, control room, MCC room, and some spares for field offices and maybe a simple field laboratory.

- Main substation and power distribution system.
- Waste treatment plants other than activated carbon adsorbers.
- Spare parts, taxes, catalyst and chemicals.
- Phase I engineering and CED monitoring.

Table 13.34 Installed Costs Historical Data

Project	Actual or Estimate	Equip. count	K$ per equipment item				Instrum. Acct.		Phase II eng. hrs/item
			1	2	3	4			
			Total cost	Equip. cost only	Phase II eng. @ $60/Hr	Commod. & field ins. (1)-(2)-(3)	TDC	K$/item	
Organics plant	A	540	201.4	34.1	45.0	122.3	No	12.7	770.
Organics plant	A	307	194.4	22.3	42.8	129.3	Yes	16.3	732.
Organics plant	A	203	171.6	19.2	27.0	125.4	Yes	14.4	462.
Organics plant retrofit	A	15	242.2	84.2	43.9	114.1	No	13.1	750.
Organics plant	A	6	295.0	58.5	107.1	129.4	No	N.A.	1,830.
Organics plant (1)	A	87	344.7	38.6	83.3	222.8	Yes	26.7	1,424.
Organics plant	A	307	205.2	60.7	50.0	94.5	Yes	15.9	855.
Organics plant	A	223	248.5	50.0	41.9	156.6	Yes	13.8	717.
Organics plant	A	118	182.9	35.9	38.8	108.2	No	14.2	663.
Organics plant	E	226	201.3	37.9	41.2	122.2	Yes	19.3	686.
Organ./mech. plant	E	57	201.8	41.5	27.1	133.2	Yes	11.7	386.
Inorg. plant	E	92	261.5	62.6	46.4	152.2	Yes	N.A.	804.
Inorg. plant	E	263	353.2	128.9	41.1	183.2	Yes	14.8	683.
Water treat.	E	134	226.0	42.8	44.6	138.6	Yes	15.6	744.

(1)This job included lethal chemicals and was built under enormous schedule pressure

Total Installed Cost

Table 13.35 is a derivation of Table 13.34 and provides very valuable historical data.

- In ten of the fourteen cases analyzed, the TIC was within 20% of $224k per equipment item with six of them within 10%.
- After backing out the equipment and engineering accounts, thirteen cases were within 20% of $130k per equipment item and 10 of them within 10%.

Table 13.35 Total Installed Costs (TIC)

	TIC k$/Item	(Accuracy Avg.) / TIC	TIC w/o Equip. & Eng. k$/Item	(Accuracy Avg. Plus Equip. & Eng.) / TIC
Actual Projects				
Organic Plant	207.4	1.08	126.0	1.02
Organic Plant	200.2	1.12	133.2	0.98
Organic Plant	176.7	1.27	129.2	1.00
Organic Plant Retrofit	249.5	0.90	117.5	1.05
Organic Plan	303.9	0.73	133.3	0.99
Organic Plant	355.0*	0.63	229.5*	0.72
Organic Plant	211.4	1.06	97.3*	1.15
Organic Plant	256.0	0.87	161.3	0.88
Organic Plant	188.4	1.19	111.4	1.10
Average	**224**		**130**	
* Excluded from avg.				
Estimates				
Organic Plant	207.3	1.08	125.9	1.02
Organic/Mech Plant	207.9	1.08	137.2	0.96
Inorganic Plant	269.3	0.84	156.8	0.90
Inorganic Plant	363.8	0.79	188.7	0.84
Water Treatment Plant	232.8	0.96	142.8	0.94
Cases within ±20%		10/14		13/14
Cases within ±10%		6/14		10/14

This information provides tools for the instant checking of the reasonableness of estimates as well the preparation of order of magnitude, or even conceptual, estimates with minimal information.

- When only the equipment count is known, an order of magnitude estimate can be prepared in seconds with the $224k factor.
- When an anotated equipment list is available, the equipment and engineering account can be estimated with the procedures discussed in this chapter, and a conceptual estimate prepared in one or two days.

An allowance must be added in both cases to account for the almost certain growth in equipment count during engineering.

The recommended allowance for first-of-a-kind plants is:

From conceptual design to final	30-40%
From Phase 0 design to final	10-20%
From Phase I design to final	5-10%

Duplicate plants should be considered as Phase 0 or Phase I.

IMPORTANT CAVEAT: These factors are very effective with plants similar to those analyzed. However, it can easily be fathomed that they would be dramatically lower for a simply instrumented plant handling low liquid flows in carbon steel equipment and requiring minimal steel structures.

Fortunately, the factors can be adjusted to broaden their scope of effectiveness with the judicious analysis of the cost breakdown shown in Table 13.36. The breakdown is based on the estimating factors of Table 5.1 after backing out the equipment and engineering costs.

REMEMBER, PROJECT MANAGERS MUST DO THE BEST THEY CAN WITH THE TOOLS THEY HAVE AT HAND.

Piping Costs

This shortcut technique is suggested for ballpark estimates of the piping account for liquid flow plants. It must be used together with Section 13.4 and confirmed with the factors in Table 5.1.

- **Process Areas**
  - Number of lines   Equipment count x 3.5
  - Length   50 ft per line
  - Number of valves   1.5 ft per line
  - Average diameter   Estimate based on pump
                                       capacity, usually 2 1/2-in.
                                       average.
  - Material   Pondered average
                                       based on plant
                                       metallurgy.

- **Interconnecting Pipe and Utility Distribution**
  Estimate based on either flowsheet and plot plans takeoffs, or use 50% of
  the process areas' piping.

Table 13.36 Field Costs Factor Breakdown

Field Activity	Thousand $		
	Material	Labor	Total
Equipment erection	0	6,300	6,300
Site preparation	300	1,600	1,900
Concrete	1,600	7,400	9,000
Structural steel	6,000	4,500	10,500
Buildings	1,900	1,200	3,100
Piping	16,000	35,100	51,100
Electrical	3,800	6,400	10,200
Instrumentation	10,900	5,100	16,000
Insulation	1,900	1,200	3,100
Painting	300	900	1,200
Fire protection	1,600	1,600	3,200
Total Directs	44,300	71,300	115,600
Field Indirects, 20% Labor			14,400
Total			130,000

Labor at approximately $45/hr.

Equipment-Related Costs

Instrumentation Account

Based on the Table 13.34 historical data, $17,000 per equipment item.

Electrical Account

An allowance of $18,000 per motor is usually more than adequate to cover:

- Unit substation and switchgear.
- Power and control wiring.
- Process area lighting.
- Grounding.

It would not cover:

- Main substation and distribution system.
- Yard lighting.
- Auxiliary buildings lighting usually included with building cost.

Engineering Account

Based on the Table 13.31 historical data, 600-700 hr per item should be adequate for the average project. Loaded engineering costs vary from $70/hr for a large contractor in the East Coast to $45/hr in the South and Southwest.

APPENDIX A
RECOMMENDED READING

- Bent, J. A. *Applied Cost and Schedule Control.* Marcel Dekker, New York, 1982.

- Clough, R. H. *Construction Contracting.* 3rd edition. John Wiley and Sons, New York, 1975.

- Guthrie, K. M. *Process Plant Estimating and Control.* Craftman Book Company of America, 1974.

- Hackney, J. W. *Control and Management of Capital Projects.* 2nd edition. McGraw-Hill, Inc., 1992.

- Kerridge, A. E. *How to Develop a Project Schedule.* Hydrocarbon Processing, January 1984.

- *Manual for Special Project Management.* A CII Publication, July 1991.

- *More Construction for the Money.* Summary Report of the Construction Industry Cost Effectiveness Project, a Business Roundtable Publication, January 1983.

APPENDIX B
GLOSSARY

AFC   Approved for construction.

AFD   Approved for design.

AFE   Authorization for expenditures.

ANSI American National Standards Institute

ASME   Americal Society of Mechanical Engineers.

Balloon   Symbol for field instrument in a P&ID.

Basic engineering   Engineering required to bring a Phase I design to the AFD level.

BW   Butt welded.

CED   Corporate Engineering Department.

CICE (Construction Industry Cost Effectiveness Project)   A construction productivity study sponsored by The Business Roundtable.

CII   An outfall of CICE.

CF   Cubic foot/feet.

CM   Construction Manager/Management.

CS   Carbon Steel.

CY   Cubic Yard.

DCS   Distributed control system.

DH (Direct hire) Practice of some general contractors of hiring craftsman directly from the local labor pools rather than subcontracting the work.

EPC (Engineering, procurement and construction) Consolidation of the responsibility for those activities under a single contract(or).

Fast tracking   Overlapping of project activities normally executed in a consecutive manner.

Flg.   Flanged.

Fringe benefits (FB)   Contractual adders to the base labor rate: medical and pension plans, vacation, travel pay, holidays, etc.

GC   General contractor.

GMP   Guaranteed maximum price.

hp   Horsepower.

HVAC   Heating, ventilating and air conditioning.

Lang Factors   Derivations of the estimating technique, originally proposed by H. J. Lang, relating total installed cost to equipment cost.

LF   Linear foot/feet.

LJ   Lap joint.

Loaded labor rate   Base labor rate plus PAC's, fringe benefits and other subcontractors costs, including overhead and profit.

MCC   Motor control center.

MPS   Master Project Schedule.

MPy   Million pounds per year.

OH   Overhead.

OSHA   Occupational Safety and Health Act.

OT   Overtime.

PAC's   Payroll added costs: Social Security, workers' compensation, insurance, federal and state taxes.

PFD   Process flow diagram. Process configuration with heat and material balances.

Phase 0   Preliminary process design.

Phase I   Firm process design.

Phase II   Detailed engineering design, procurement and project control.

P&ID   Process and instrumentation diagram. Basics of detailed engineering.

PLC   Programmed logic controller.

PO   Purchase order.

PM/PE   Project Manager/Engineer.

QA/QC   Quality assurance and control.

SC   Subcontract/subcontractor.

SF   Square foot/feet.

SS   Stainless Steel.

SW   Socket weld.

Swd   Screwed.

Sy   Square yard.

Take-off   Detailed quantity count of work components: cubic yards, tons, feet, etc.

TEFC   Totally enclosed fan cooled. Term applied to electric motor.

T-T   Tangent to tangent. Straight-side dimension of vessels, columns, reactors.

VM   Venture Manager.

WU   Work Unit. Standard unit established to value all work components in a rational and consistent manner.

APPENDIX C
TYPICAL COORDINATION PROCEDURE

Introduction

The _____ Division had approved pre-AFE spending to begin process design for a _____ production facility at the _____ plant site. Final AFE approval is expected by _____ and _____ production should start no later than _____ in order to meet marketing requirements.

Basic Guidelines

The facility should be designed for a _____ to _____ year life. The economics of the product indicate that low capital cost is (is not) more important than low operating costs.

The most cost-effective schedule will be determined during AFE preparation but all schedule improving measures that could have an adverse effect on capital cost must have specific management approval.

In order to minimize cost through the most effective use of in-house personnel, we will follow the small project approach acting as general contractor and subcontract engineering and construction services on a discrete basis as required.

General Responsibilities

_____ Division

The _____ Division (plant) is the project sponsor, owner and eventual operator of the facility. They provide overall direction as well as design criteria, basic process information and operating experience and have the ultimate responsibility for the success of the project.

Coordination and overall direction of the project comes form the Division (plant) through the Venture Manager.

Fundamental process data, as well as operating and maintenance experience, come from the plant and/or R&D through the various specialists assigned to the design team.

The plant will provide P&ID drafting, purchasing and accounting services as well as a field manager to direct construction activities.

Central Engineering Department

CED is responsible for the design and construction of the facility within the scope established in the AFE. The process design (Phase I) will be done by the CED process section with cooperation from Division R&D and plant engineers.

The task force or team concept prevails. All personnel are responsible to the Project Manager when acting as members of the project team and/or assisting the Project Manager in performing his/her duties with respect to the project.

CED will also do all contracting, obtain prices, and prepare requisitions for the construction materials purchased by the plant.

Corporate Environmental Planning

The Corporate Environmental Planning Department is kept informed of environmental issues and problems. The Project Manager and design team seek advice and assistance from these groups as specific needs arise generally through the plant Environmental Department.

Safety and Industrial Hygiene

The Group Safety and Industrial Hygiene Manager is kept informed of hazards problems by the Venture Manager through the plant safety supervisor. The Project Manager and design team seek advice, recommendations and assistance from this resource, including coordination of safety audits.

Corporate Health Services

The Health Services group is kept informed of hazards problems by the Venture Manager. The Project Manager and design team seek advice, recommendations and assistance from this resource.

Individual Responsibilities

Venture Manager (Name)

The Venture Manager reports to (Position)

- Represents the business group and is responsible for the establishment of project financial and schedule goals and guidelines.
    - Evaluates developmental engineering efforts coordinating business objectives with realistic execution programs.
    - Reviews business and technological alternatives available and makes a recommendation to business group management.
    - Coordinates venture efforts with corporate groups, marketing, production, purchasing, legal, etc.
    - Assembles the project design team.

- Provides the overall direction and assumes the responsibility of coordinating the efforts of all internal groups and insuring that operable, safe and economic facilities are designed, built and started up.
    - Secures all permits, licenses or other certificates required by governmental agencies.
    - Makes all business arrangements directly associated with the project except those related to contracting engineering and/or construction services.
    - Maintains liaison with Environmental Planning, Health Services, Toxicology and other corporate staff departments.
    - Ascertains that the Project Team receives adequate and timely input and support form R&D and Operations.
    - Organizes and supervises the start-up operations.
    - Keeps Division management informed and secures approvals as required.

Project Manager (Name)

The Project Manager has a dual responsibility to both the Venture Manager and to the CED Director for the execution of the engineering, procurement and construction.

The Project Manager is the official CED contact with the contractors and has the ultimate responsibility of directing and controlling the scope, budget and schedule as defined in the AFE. Through the Technical Manager and the Field Manager, the Project Manager manages and coordinates the activities of the design team, plant support, the contractors, other agencies and the staff specialists assigned to the project. Duties include but are not limited to:

- Support the Venture Manager during the early stage of project development.
    - Develop project execution strategy and conceptual schedules to ascertain the viability of the Venture Manager's objectives.
    - Prepare the preliminary project execution plan.
    - Assume a hands-on participation in the preparation, review and approval of the AFE estimate.

- Support the Technical Manager during the preparation of Phase 0/I design packages.
    - Prepare conceptual estimates of the process options being considered to assess their cost impact.
    - Provide project engineering and constructability input to plot plans and equipment arrangement options.

- Monitor, approve and document all project expenditures, changes and charges in order to insure proper control.
    - Review and approve, within the established limits of authority, all vendors' and subcontractors' quotations and bid analyses.
    - Review and approve all change orders and claims to confirm their validity, scope, cost, proper approvals, documentation, timeliness, etc.
    - Review and approve all invoices prior to payment.
    - Approve all internal charges to the AFE to ascertain their correctness.

- Direct, supervise and coordinate the work of all members of the Project Team as well as the various internal supporting groups.
    - Develop the project participants into a team, with a team attitude and focus, so that all participants are free to contribute and participate to the fullest and most beneficial extent.
    - Insure that the owner's reviews and approvals are conducted in a timely manner.
    - Insure the timely participation of the staff specialists.
    - Keep the various support groups informed so that they can provide their input in a timely manner.
    - Review and approve Phase 0/I packages for completeness before transfer to the engineering contractor.

- Assume the responsibility for the mechanical design and construction of the AFE scope assigned to CED, to achieve the quality cost and schedule objectives stated in the AFE.
  - Prepare the project execution plan.
  - Prepare the technical bid package, develop bidders qualifications and contractor evaluation criteria. Lead the technical evaluation effort.
  - Direct and supervise the work of the engineering contractor, monitoring its procedures, staffing and performance to insure that the quality, cost and schedule objectives are met without unnecessary expense and within timing constraints.
  - Review and approve contractors' schedules and cost estimates.
  - Review, on a routine basis, the contractors' progress and establish control systems to make independent evaluations.

- Maintain proper project records and documentation.
  - Organize and maintain proper project files.
  - Keep up-to-date records of all changes and extras.
  - Keep up-to-date records of all commitments and expenditures.
  - Ascertain that all important project decisions are formally documented through minutes of meetings, memos and/or formal change notices.
  - Insure that contractors submit all the required documentation, schedules, reports, insurance policies, etc., on a timely basis.

- Keep the Venture Manager and CED management informed of all significant developments, particularly those concerning changes, cost and completion dates.
  - Issue monthly status and cost reports.
  - Make monthly schedule and cost forecasts.
  - Prepare close-out report.

Technical Manager (Name)

During the process design stage, the Technical Manager is responsible to the Venture Manager and during the execution stage, to the Project Manager. Duties include:

- Assume the responsibility for the technical contents of the design.

- Develop an adequate and economical process design (Phase 0) that is consistent with the design, operating and maintenance criteria established by the Venture Manager.
- Develop the detailed process and engineering specifications and P&ID's (Phase I) required to implement the approved scope.
- Work closely with the Venture Manager, R&D, plant process engineers and, if required, outside consultants to achieve consensus on best process choices and insure that proper and sufficient design data are developed on a timely basis to support the Phase II engineering.
- Direct the work of the Process Engineers assigned to the Project Team.
- Conduct hazards and safety reviews in cooperation with the appropriate personnel and, if required, outside consultants to insure a safe, efficient plant start-up and operation.

- Insure contractor's compliance with approved design specification and scope.
  - Transmit all technical information to the contractor.
  - Review and approve contractors' technical work, process calculations, equipment layouts, project specifications and, above all, any changes to the P&ID's.
  - Review and approve technical content of vendors' quotations and bid summaries.
  - Document all technical decisions and/or changes made by the owner and/or the contractor.
  - Interface with internal and contractors' specialists to obtain their input when required.
  - Prepare design manuals.

Construction Manager (Name)

The Construction Manager has a dual responsibility to both the Project Manager and the plant engineer. The Construction Manager represents the Project Manager in the field and is the official contact with construction contractors in all field-related matters. Generally, the CM is responsible for all field activities including those performed directly by the Owner's personnel. The duties include but are not limited to:

- Direct and/or coordinate both contractors' and plant's field work to insure smooth interfacing.

- Monitor, through the field inspectors, all work to insure conformance to specifications, schedule, safety and plant regulations.
- Direct and supervise the field inspectors and the project safety engineer.
- Assure proper plant security and safety in all areas related to the AFE work.
- Review and approve all field extras within limits of authorization.
- Coordinate contract work with other plant activities.
- Review and approve detailed schedules of any construction work affecting plant operations.
- Make independent progress evaluations and projections.
- Write field progress reports to keep Project and Venture Managers informed.
- Monitor the work performed by the plant to ascertain whether it is done on a timely basis, and advise contractor and Project Manager of any schedule deviations.

Process Engineers (Names)

Process Engineers have a dual responsibility to both the Technical Manager and their line supervisors. Their responsibilities are to:

- Develop an adequate and economical process design and control criteria consistent with safety, operating efficiency and maintenance economy.
- Develop, review and approve process flowsheets and P&ID's.
- Review the instrument engineer's work to ascertain compliance with the established control criteria.
- Document all important technical decisions.
- Conduct operating and hazards reviews.
- Prepare operating procedures.
- Conduct final check-out and assist the production group during plant start-up.

Production Engineer (Name)

The Production Engineer is in charge of the start-up and operation of the facility. In that capacity, the PE has a dual responsibility to both the Venture Manager and the unit production manager.

CED/Division Specialists as required (Names)

Specialists have a dual responsibility to both the Project Manager and their line supervisors and will support the project work in their discipline areas as required by the Project Manager. These areas include:

- Auditor
- Contract
- Cost
- Electrical
- Hazards
- Instrumentation
- Mechanical
- Procurement

Limits of Authority

Purchasing and Subcontracting

All commitments and expenditures, including work performed by the plant, will be covered by purchase order and/or plant work orders which must be authorized by the Project Manager or a designate. Unauthorized charges will not be honored.

Approval levels for purchase orders and subcontracts within budget are as follows:

Up to $_____	Project Manager
$_____ to $_____	Venture Manager
Over $_____	Venture Manager CED Director

Design/Scope Changes

After approval of the AFE, deviations from the scope as defined in the AFE and/or an approved design shall be handled in a formal manner and subject to the following levels of approval:

Up to $_____	Project Manager
$_____ to $_____	Venture Manager

$_____ to $_____   CED Director

$_____ to $_____   Manufacturing Director

$_____ to $_____   Group Vice President

$_____ and over   President

Field Changes/Extras

After award of each contract, all field changes and extras shall be documented in a timely manner according to the procedure set up in the contracts.
The authorization levels required have been set as follows:

Up to $_____   Construction Manager

$_____ to $_____   Project Manager
                              Venture Manager

$_____ and over   To be handled as Scope Changes

APPENDIX D
ESTIMATE CHECKLIST

Purpose

To call to mind various items required in the preparation of an estimate that should be considered for inclusion.

- General:
  - Contract type:
    1. Lump sum.
    2. Cost plus with/without incentives.
  - Environmental:
    1. Permits - associated delays.
    2. Special pollution control requirements.
    3. Environmental Impact Statement.
  - Weather conditions - seasonal effects:
    1. Winterization.
    2. Rain.
  - Job conditions
    1. Congested work area accessibility.
    2. New technology.
    3. Special safety requirements.
    4. Complex process - highly instrumented?
    5. Revamp of existing facilities.
    6. Utility tie-ins.
    7. Interferences due to operation? Productivity?
    8. Scope well defined? Off-sites?
    9. Plot plan, flow sheets.
    10. Schedule - manpower loading, availability.

11. Local construction climate - overtime required?
12. Union/non-union.
13. Special fabrication requirements:
    a. Stress relief, X-ray.
    b. Hydrostatic testing.
    c. Pneumatic testing.
14. Special paints.
15. Insulation.

- Site Development:
  - Land purchase.
  - Drainage, dewatering.
  - Cut and fill.
  - Removal of excavated material - contaminated wastes, spoils.
  - Soil stability.
  - Landscaping, seeding, topsoil, fertilizer.
  - Clearing and grubbing.
  - Spillways.
  - Signs.
  - Bridges.
  - Roadways - concrete, asphalt, gravel, other.
  - Walkways - concrete, asphalt, gravel, other.
  - Fencing - property, security chain link, other.
  - Railroads - siding, switching, bumpers.
  - Sewers - storm, sanitary, acid, other.
  - Area lighting.
  - Tank dikes - earth, concrete, other.
  - Special transportation to remote sites.
  - Labor camps, community facilities.
  - Underground obstructions.
  - Existing pipe racks adequate?
  - Piling - bearing, sheet.
  - Parking facilities.
  - Guard service.
  - Housing accommodations.

- Equipment:
    - Spare equipment - in place, warehouse spares.
    - Truck and/or rail scales.
    - Firewater storage tank, fire water pump, emergency power - diesel, electric.
    - Truck loading/unloading facilities.
    - Rail loading/unloading facilities.
    - Barge loading/unloading facilities, docks.
    - Forklift trucks.
    - Hauling vehicles, trains, trucks, tractor.
    - Ambulance.
    - Firetrucks.
    - Vendor representatives, field service start-up assistance.
    - Start-up.
    - Grouting, setting, acid brick.
    - Safety equipment.
    - Freight - domestic, overseas, air.

- Utilities:
    - Water - city, wells, seawater.
    - Water treatment.
    - Process water treatment.
    - Process waste water treatment.
    - Process waste treatment - outfall treatment or monitoring.
    - Boiler - cogeneration.
    - Boiler - single fuel, multifuel.
    - Boiler - feed water treatment.
    - Boiler - high and low pressure steam and condensate return lines.
    - Compressors - plant air.
    - Compressors - instrument air, receiver, dryer and main header.
    - Compressors - process gas.
    - Emergency power - electric/diesel generators, inert gas.

- Firewater systems - tanks, pumps, underground pipe.
- Sprinkler systems - wet/dry, heads, hose cabinets.
- Cooling tower, cooling tower pumps, header piping.
- Air venting systems - blower, duct, connections supports.
- Environmental - air purifying, water, dust collecting.
- Metering systems.
- Stacks.

- Buildings - type, style, finish, openings:
  - Process building.
  - Warehouse.
  - Shop, shop crane/hoist.
  - Administration and office buildings, office furniture, equipment.
  - Laboratory, lab equipment computer, analyzers, other equipment.
  - Change house, restrooms, shower, lockers (male and female), lunch room.
  - Control building - blastproof, computer system.
  - Motor control center.
  - Substation.
  - Compressor building.
  - Garage.
  - Pump house.
  - Guard house.
  - Structures - open, closed.
  - Partially enclosed storage areas.

- Miscellaneous:
  - Pipe racks - bents, supports, hangers, sleepers, t-supports.
  - Piping.
  - Instrumentation, instrument pipe, electronic, digital, pneumatic.
  - Automation - process controllers, host computers, robotics.
  - Electrical - power wiring, MCC, substation, X-former, instrument wiring, grounding poles, pole lines, lightning arrestors.
  - Concrete.

- Steel.
- Fireproofing.
- Painting.
- Insulation.
- Field costs:
    1. Base rate - journeyman.
    2. Fringes - vacation, union dues, pension fund, medical coverage, travel, other.
    3. Payroll taxes.
    4. Payroll insurance.
    5. Consumable supplies.
    6. Petty tools.
    7. Temporary facilities - trailers, water, toilets, change house, parking.
    8. Unallocated labor.
    9. Construction equipment rental.
    10. Overhead.
    11. Profit.
- Contract home office and fee.
- Consultants.
- Taxes - federal, state, local.
- Duct work, supports, stiffeners, louvers, diffusers.
- Heating, ventilating and air conditioning.
- Elevators - freight, passenger.
- Handicapped facilities.
- Protective equipment.
- Expense items - dismantling, repair, rework, removal.
- Taxes - special taxes, duties, imports.
- Spare parts.
- Start-up.
- Escalation.
- Restricted reserve.
- Contingency.

APPENDIX E
TECHNICAL EVALUATION CRITERIA
EXAMPLE

The following example has been taken from the file of an actual project. It illustrates the procedure followed for the final selection of a contractor from a list of bidders who had been qualified previously.

The evaluation criteria and guidelines presented in Tables E.1 and E.2 were prepared by the Project Manager with input from process and cost engineering.

General Guidelines

- At this point, all bidders are technically acceptable and we are only trying to compare them with each other.
- Rate each criteria item from 1 to 10.
- The best should be rated 10.
- If possible, avoid ties. If two or three are considered to be very close to each other, use a narrow spread, e.g., 10.0, 9.8, 9.5.
- For some of the criteria, it could be practical to have each evaluator do an independent rating before comparing the results. In other cases, it could be more practical to do the rating jointly.
- In all cases, the final result must reflect the consensus of the evaluation team.
- If we are required to come up with a preliminary rating based on the proposal only (before visits), the rating must identify those criteria items when the rating could change as a result of the visit. The impact of the potential changes in the final rating must be evaluated.

Table E.1 Evaluation Criteria

Criteria	Points	
General		
Project approach	6	
Construction approach	5	
Controls	5	
Management commitment	4	
Flexibility	3	
Procurement	3	
Size of projects	2	
Size of company	1	
Stability	1	
Subtotal		**30**
Specific to Project		
Process		
Relevant experience of proposed team	7	
Experience with first-of-a-kind process	5	
Availability and use of computer design area	5	
Availability of qualified personnel	5	
Process management commitment	3	
Key Personnel		
Project manager	5	
Experience together	5	
Process leader	4	
Construction manager	4	
Cost engineer (planner)	3	
Subcontract administrator	2	
Instrument engineer	2	
Instrumentation and control	5	
Other disciplines	5	
Understanding of work	5	
Experience in the plant area	5	
Subtotal		**70**
GRAND TOTAL		**100**

Table E.2 Evaluation Criteria Guideline

Criteria	Positive	Negative
General		
Project approach 6 points	Full & real responsibility of P.M. throughout project	Split of responsibility from eng. to const.
	Task force (modified) approach	Matrix approach
	Understanding that S.C. work requires different planning than D.H. work	Trying to prove that D.H. is the best approach to projects
Approach to construction 5 points	Proven capability to perform both in S.C. or D.H. situation	Usually works on D.H. basis
	Objectivity in discussing D.H. vs S.C.	Can't do D.H.
		Proposes very heavy field staff, > 5% craft labor
Controls planning/ estimating costs 5 points	Prominent position in organization	System so elaborate & detailed that reporting is delayed
	Separate line of reporting	
	Hands-on participation	
	Conduct regular audits (H.O./field)	
	Computerized systems	
	Regular cost tracking mtgs	
Management commitment 4 points	Personal interest shown by high executive	Company has substantial backlog
	Personal relationships	Very large jobs going
Flexibility 3 points	P.M. has management support to "bend" rules	Many managerial & approval levels
	Client's P.M. has access to "raw data"	Emphasis on formal communications
	Direct communication at working levels	
Procurement 3 points	Have separate inspection/ expediting	Tech personnel have negative attitudes towards purchasing department
	Will have dedicated coordinator in task force	
	Willingness to be flexible	

Table E.2 (Continued)

Criteria	Positive	Negative
Size of projects 2 points	20-50M	Our job less than 5% of capacity
Size 1 point	400-600	
Stability 1 point	High average seniority	Recent layoff Frequent reshuffling of personnel

Specific to Project

Process

• Relevant experience of proposed team 7 points	Depth of experience in liquid phase stirred reactions – liquid-liquid extraction & non-ideal distillation	Out-of-date with latest technology
	Avg. experience, 5-15 years	Less than 3 years experience
	Familiarity with computer design aids available in firm	
• Experience with first-of-a-kind designs 5 points	Company has experience Individuals have experience	
• Availability & use of computer design aids 5 points	ASPEN, in-house non-ideal distillation programs, etc., available & regularly used	
• Availability of qualified personnel 5 points	6-8 experienced process engineers	
• Process management commitment 3 points	Have special unit for development work Manager of process eng. will have active involvement in project	

Table E.2 (Continued)

Criteria	Positive	Negative
Key Personnel		
• Project manager 5 points	Several project of similar nature	First or second project with firm
	Subcontract experience	> 20 yrs with same company
	Field experience	Specialization in any discipline
	General engineering experience	Relies too much on discipline managers
	Strong commitment to project/ client	
	Planning experience	
• Experience together 5 points	Have succesfully worked together in previous job	
• Process leader 4 points	Has first-of-a-kind process experience	First or second project with firm
	Previous experience as process leader	
	Have expertise in areas specific to our project	
• Construction manager 4 points	Engineering degree	Previous involvement with government work
	Planning experience	First or second project with firm
	Experience with subcontractors	
	Understands role of planning & cost control & uses them	
• Control engineer (planner) 3 points	Project engineering experience	First or second project with firm
	Field experience	
	Previous experience as lead man in other jobs	
• Subcontract administrator 3 points	Engineering/purchasing &/or legal background	
	Familiar with area subcontractors	

APPENDIX F
IN-HOUSE CONSTRUCTION PROGRESS
MONITORING SYSTEM EXAMPLE

Scope

A project estimated at $1.0M and 10,000 construction hours with the following major cost components:

Site Work

Estimated at $250k and roughly 1000 hr. Must be done during the rainy season and includes cleaning up an old chemical waste dump. The cost and schedule overrun risks are very high and no other work can proceed until all site work is completed.

Foundations

Estimated at $100k and 3,000 hr. Suspected presence of underground boulders and some old foundations. High potential for overrun. Building and mechanical work can start but cannot be completed before all foundations.

Building

Estimated at $150k and 1,500 hr. Pre-engineered building with metal siding. No problems anticipated. Mechanical work can start but not be completed before building is completed.

Mechanical Work

Estimated at $500k and 4,500 hr. Installation of a few pumps and tanks and approximately 4000 feet of mostly straight alloy pipe. Minimal overrun potential.

Relative Values

The relative value assigned to each of the above components could be based on the dollar cost, estimated hours or on a pondered basis that considers both as well as the potential cost and schedule exposure.

	Cost Basis		Hr Basis		Pondered Basis	
	$k	Fraction	Hr	Fraction	Exposure	Fraction
Site work	250	0.25	1000	0.10	Very high	0.25
Concrete	100	0.10	3000	0.30	High	0.25
Building	150	0.15	1500	0.15	Normal	0.15
Electr./mech.	500	0.50	4500	0.45	Low	0.35
TOTAL	1000	1.00	10000	1.00		1.00

NOTE: The pondered basis fraction is obtained based on a personal judgment considering the dollar cost, man-hours involved and risk exposure. When these factors are considered for each major component a pondered basis fraction is assigned.

Considering the schedule importance and potential risk involved in the site preparation work and the simplicity of the mechanical work, the pondered relative values would provide a more meaningful parameter to gauge progress.

The total value of the work is now unity and will remain unchanged regardless of scope addition or subtraction. The major components as well as the comprised related construction activities are expressed as fractions.

Each of the major components must be broken down into discrete activities consistent with the cost estimate breakdown, the execution plan and the nature of the activity. The relative value of each activity within a given major component must be assigned on the basis of either cost, man-hours and/or physical units (yd^3, ton, ft, ft^2, etc.) and prorated to the component fraction.

Site Work (Fraction: 0.250)

Since the activities involved are very diverse and the estimated hours are not a true reflection of their real values, the assigned fractions are based on the pondered assessment of cost and man-hours.

Activity	Qty	Cost basis		Mhr Basis		Pondered Basis	
		$K	Frac.	Mhr	Frac.	Site Frac.	Total Frac.
Excavation	500 yd^3	10	0.04	100	0.10	0.10	0.025
Disposal	300 yd^3	150	0.60	100	0.10	0.30	0.075
Backfill	200 yd^3	40	0.16	200	0.20	0.20	0.050
Sewers	500 LF	20	0.08	300	0.30	0.20	0.050
Paving	1000 yd^2	30	0.12	300	0.30	0.20	0.050
TOTAL	N.A.	250	1.00	1000	1.00	1.00	0.250

The progress of each activity can be gauged in the most practical manner for each one.

- Excavation, disposal and backfill by cubic yard and/or truck load.
- Sewers by line and/or linear feet.
- Paving by square yards.

Concrete Work (Fraction: 0.250)

The activities involved are straightforward and their value can be related to work hours:

Activity	Yd^3	Hr	Concr. Frac.	Total Frac.
Building foundations	100	1500	0.50	0.125
Building grade beams	20	450	0.15	0.038
Building floor slab	50	300	0.10	0.025
Area paving	45	300	0.10	0.025
Pump pads	10	150	0.05	0.012
5 Tank foundations	25	300	0.10	0.025
TOTAL	250	3000	1.00	0.250

NOTE: If so desired, the building foundations could be broken down in several sections.

The progress of each activity can be gauged with a milestone system.

Building (Fraction: 0.150)

If the estimate does not include an activity breakdown, the project manager uses experience and/or imagination to break the work into the conventional building construction activities.

Activity	Bldg. Frac.	Total Frac.	Progress Gauging
Main structure	0.10	0.015	Eyeballing
Roofing	0.15	0.022	ft^2
Siding	0.15	0.023	ft^2
Doors & windows	0.10	0.015	Unit
Lighting	0.20	0.030	Milestones
HVAC	0.20	0.030	See note
Finish	0.10	0.015	Eyeballing
TOTAL	1.00	0.150	

NOTE: HVAC work could be broken down by power supply, equipment installation or ducts installation.

Electrical/Mechanical Work (Fraction: 0.350)

The activities involved are very diverse:

- Equipment erection.
- Miscellaneous structures.
- Piping fabrication and erection.
- Power wiring and lighting.
- Instrument wiring.
- Insulation.

Most of these could be dissected into small components, all of which are related to work-hours and can be gauged with a milestone system. The hours are easily prorated to the 0.350 total fraction.

IMPORTANT NOTE: TO AVOID POTENTIAL MISTAKES WITH THE USE OF DECIMAL FRACTIONS, THEY SHOULD BE MULTIPLIED BY 1000.

APPENDIX G
FORECASTING FINAL SUBCONTRACT COST

Introduction

When a project is executed through a competitive lump sum subcontract, the Project Manager must frequently take a calculated risk in order to maintain the schedule and start awarding subcontracts with partially complete detailed engineering. The result is an almost certain increase in cost between subcontract award and completion.

The increase in cost must be anticipated by adding, upon award, an adequate resolution allowance to the base subcontract cost. This resolution allowance must be reevaluated periodically as the work progresses and new informaiton becomes available.

The following tools are required in order to exercise project cost control and provide accurate forecasts:

- Consistent definitions of the various subcontract related costs.
- A method to evaluate and assign, in a rational manner, the resolution allowance required for each subcontract.
- A method to reevaluate and update the resolution allowance as work progresses.

This guideline addresses all three requirements and is intended mainly for electromechanical work.

Subcontract Cost Definitions

Base Scope

Work covered by the bid drawings and specifications agreed upon at the time of subcontract award. The award cost must be treated as "committed". Items on "hold", even if related to the initial scope, are not considered part of the initial scope. However, they must be identified and treated as "uncommitted" at their current estimated cost and included in the cost forecast.

Additional Scope

Work resulting from the resolution of "holds" and/or major design changes in the initial scope. When defined and added to the subcontract, they must be considered as "committed" at the subcontractor firm price or best estimate, as applicable. Changes in take-off quantities arbitrarily fixed for bidding purposes will fall in this category. The "committed" quantity must be adjusted based on unit prices as soon as the real quantities are known.

   Correction of oversights and miscellaneous changes resulting from detailed design are not considered additonal scope. They must be considered subcontract changes, as indicated below.

New Scope

Work unrelated to the initial scope that is added to the subcontract. When incorporated into the subcontract, it must be treated as "committed" at the subcontractor's firm price or best estimate, as applicable.

   New scope to a subcontract could be either work already identified in the AFE estimate or work resulting from scope changes.

Base Subcontract Cost (BSCC)

The BSCC is the baseline for cost tracking and determination of the resolution allowance. It is the cost of the base scope plus any additional scope and/or new scope as applicable. It does not include changes. The BSCC could be:

- A lump sum developed by the bidders (subcontractors) based on the bidding documents.
- A cost derived from unit price developed by the bidders applied to approximate materials take-offs provided by the client.

Subcontract Changes (Growth)

Generally, subcontract changes would come from:

- Engineering (home office) design changes:
    1. Changes developed through detailed engineering.
    2. Minor P&ID revisions.

    3. Scope related items not identified at subcontract award.
    4. Correction of errors and oversights, regardless of size, in the
        original subcontract documents.
-   Field changes:
    1. Corrections of errors made by engineering and/or vendors.
    2. Additional work and or changes requested by client in the field.
    3. Cost variances in items incorporated in the subcontract on an
        estimated rather than lump sum basis.
    4. Shortages of materials or services that were to be supplied by
        others.
    5. Extra cost resulting from unreasonable delays in delivery or
        materials and/or engineering information.
    6. Extra costs resulting from delays related to plant operation needs not
        covered in the bid package.
-   Punchout/precommissioning:
    1. Miscellaneous last-minute work not covered by subcontract scope
        and/or changes required for start-up of the facility.

Resolution Allowance

The resolution allowance is intended to cover the "normal growth" of the BSCC
that reflects subcontract changes as defined above. Since the resolution allowance
represents an expected cost, it is not treated as contingency but must be included in
the cost forecast and shown initially as "uncommitted" and gradually transferred to
"committed" as the changes materialize. Major design changes are covered by the
contingency (refer to Section 13.10 for a better understanding of contingency and
resolution allowance).

Resolution Allowance Criteria

The "normal growth" occurs gradually through three well-defined execution phases
and is due to:
-   Design changes (engineering phase).
-   Field changes (construction phase).
-   Punchout items (commissioning phase).

The extent of each depends on the circumstances particular to each project:

-   Schedule: normal/fast track.
-   Type of plant: Grass roots/retrofit.
-   Subcontractor size.

- Depth of client's review.

All the preceding factors must be taken into consideration when the resolution allowance is set for each subcontract.

Table G.1 proposes a criteria to do so.

Monitoring Subcontract Growth and Updating Resolution Allowance

- As the work progresses, the subcontract cost will grow as a result of design and field changes.

  For good cost forecasting, the growth must be monitored and projections made to ascertain that the final growth will be consistent with the resolution allowance. If not, the resolution allowance must be adjusted as required.

  The correlation in Figure G.1 was developed from three subcontracts, two piping/mechanical, and one electrical/instrumentation. The BSCC of each one was approximately $1 million. Even if the growth factor were different in each case, the rate of growth followed a very consistent pattern. This correlation should provide a good tool for monitoring growth and adjusting the resolution allowance when so required.

- This correlation should be most effective when:

  - Both design and field changes are recognized and incorporated to the cost on a timely manner even if based on preliminary estimates. In this case, a poor estimate is much better than no estimate.
  - The field supervision has established an accurate progress measurement system independent from the subcontractors'. If this is not available, subcontractors' progress reports can be used. In this case, it must be remembered that most subcontractors will over-evaluate progress by 5-10%.

The portion of the resolution allowance allocated to punchout/precommissioning (see Table G.1) should be left intact until the work is at least 95% complete.

Table G.1 Initial Resolution Allowance Criteria

| Type of Change | Criteria for Minimum Changes | Resolution, % of BSCC | | |
		Min.	Normal	Max.
Engineering	• Minimum schedule pressure			
Design Changes	• AFC basic drawings			
	– P&ID's			
	– Arrangement dwgs			
	– Piping dwgs	2	2 - 4	6
	• Detailed engineering			
	– Isometrics / schematics			
	– Steam tracing details			
	– Piping supports			
Field Changes	• Minimum schedule pressure			
	– Complete bid package			
	– Intelligent bids			
	– Thorough negotiations			
	• Grass roots facility			
	– New equipment	4	7	12
	• Extensive review/approvals			
	• Model			
	• Plant commitment			
	• Subcontractor			
	– Sophisticated			
	– Conservative			
Punchout /	• Extensive reviews/approvals			
Precommissioning	• Model	2	4	7
	• Plant commitment/V.M.			
TOTAL		8	15	25

Type of work	BSCC. k$	Growth, k$	Final cost, k$	% Growth
• Equip./pipe	1,100	330	1,430	30.0
x Equip./Pipe	1,085	165	1,250	15.2
+ Elect./Inst.	1,072	168	1,240	15.7

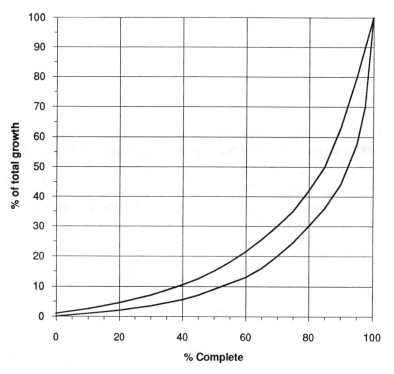

Figure G.1 Rate of subcontract growth.

APPENDIX H
HEAT-TRACING MODELS

Electric Tracing

Winterizing Service

- One 250-ft 50-amp, 480-v feeder.
- One 480/120-v 15-kva transformer.
- One distribution panel with ten 20-amp 120-v switches, 8 active and 2 spares.
- Eight 40-ft 120-v lead wires.
- Eight 200-ft runs of 5 BTV 1-ct heater tape, complete with thermostat, power connection kit, end seals and contactor.

Average actual pipe coverage 160 ft.

250°F Process Service

- One 250-ft 100-amp 480-v feeder.
- One 480/120-v 40-kva transformer.
- One distribution panel with ten 40-amp 120-v switches, 8 active and 2 spares.
- Eight 40-ft 120-v lead wires.
- Eight 160-ft runs of 15 XTV 1-ct heater tape, complete with thermostat, power connection kit, end seals and contactors.

Average actual pipe coverage 130 ft.

300°F Process Service

- One 250-ft 200-amp 480-v feeder.
- One 480/240-v 75-kva transformer.
- One distribution panel with ten 50-amp 240-v switches, 8 active and 2 spaces.
- Eight 40-ft 240-v lead wires.
- Eight 280-ft runs of 20 KTV 11-ct heater tape, complete with thermostat, power connection kit, end seals and contactors.

Average actual pipe coverage 220 ft.

Steam Tracing

- One 30-ft long 2-inch diameter insulated steam feeder with filter and self contained ambiant temperature control valve.
- One 2-inch diameter insulated steam manifold complete with filter and trap assembly. Six 1/2-inch valved connections and two capped spaces.
- Six 40-ft runs of preinsulated 1/2-inch steam tubing.
- Six 75-ft runs of 1/2-inch copper tubing.

Average actual pipe coverage 60 ft.

- Six 40-ft runs of preinsulated 1/2-inch condensate tubing.
- One 2-inch diameter insulated condensate manifold with six 1/2-inch connections, each with a filter and trap assembly and two capped spaces.
- One 20-ft 2-inch diameter insulated, valved condensate line.

APPENDIX I
FIELD INDIRECTS CHECKLIST

Total Cost

The total cost of the field indirects is 100-130% of direct labor costs (direct hire) plus 15-20% of subcontracts and inversely proportional to project size. In the following breakdown, an asterisk denotes services usually supplied by Contruction Manager or General Contractor.

	Percent of Direct Hired Labor Costs
Labor	
Indirect Craft Labor	*15-20*

1. General foremen
2. Temporary facilities builders*
3. Temporary facilities maintenance*
4. Chainmen on layout crew*
5. Janitors*
6. General cleanup crews*
7. Drivers (non-construction)
8. Tool room attendants
9. Warehouse helpers*
10. Fire watch
11. Equipment and weather protection
12. Show time
13. Evacuation alarms and tests*
14. Sign-up/termination time
15. Standby time
16. Craft training
17. Safety meetings
18. Security checks

Indirect Non-Craft Labor 5-6

1. Secretaries
2. Clerical/accounting/cost/engineering
3. Timekeepers
4. Guards/watchmen
5. Nurses
6. Storeroom attendants
7. Draftsmen

Labor PAC's Craft and Non-Craft 20-25

1. FICA
2. Workmen's Compensation
3. Federal/state unemployment taxes

Craft Fringe Benefits 20-25

1. Medical/pension plans
2. Holidays/vacations
3. Travel/subsistence pay

Materials

Temporary Facilities (Build and Maintain) 5-7

1. Offices*
2. Warehouses*
3. Tool rooms
4. Change rooms
5. Sanitary facilites*
6. First aid room*
7. Fabrication shops
8. Instrument calibration shop
9. Guard house/brass alley*
10. Roads/fences/parking/drainage/signs/lighting*
11. Equipment and weather protection shelters
12. Office furniture/computers*
13. Temporary utilities*
14. Laydown and storage areas*
15. Repair plant roads*

Supplies*   *2-4*

   1. Office supplies
   2. Reproduction
   3. Communicaitons
   4. Medical supplies
   5. Janitorial supplies

Consumables   *4-5*

Small Tools   *4-5*

Equipment Rentals

Equipment   *8-12*

Fuels and Lubricants   *3-5*

Home Office Supervision*

Direct Field Supervision   *5-7*

Administrative Support   *2-4*

Travel Expenses   *per contract*

Other

Contractor's Field Overhead   *per contract*

Premium Pay   *4-6*

INDEX

Agency agreements, 84
Anatomy:
  of estimate, 65-70
  project control system, 156-158
Audit exceptions, 181-183

Bid analysis/evaluation, 94, 101
Bidders qualification/selection, 93-94,
  99-100
Bidding, 94, 100-101
  instructions, 98-99
Bid package, 98-99
  preparation of, 90-91
Building:
  account, 76
  estimating, 219

Case study, 16-21, 41
Checking:
  contractors schedule/execution
    plan, 171-173
  cost estimates, 79-82
Civil work estimating procedure,
  208-220
Coemploymentship, 140
Columns estimating (see Vessels
  estimating)

Communications, 7, 184-187
  criteria and guidelines, 184-185
  documentation checklist, 185-187
Competitive contract, 85
Compressing the schedule, 50- 53
Conceptual:
  design, 23-25
  estimates, 58-59
  estimating, 274-279
Concrete:
  account, 75-76
  estimating procedure, 208-214
  takeoff guidelines, 209-214
  unit prices and work hours, 208
Conceptual plant layout guidelines
  (see Plant layout, guidelines)
Construction industry cost effectiveness report (CICE), 139
Construction industry institute, 139
Construction management, 129-139
  CICE influence on, 139
  Construction manager (C.M.) acti-
    vities, 131-134
  construction options, 130-131
  overview, 129-130
  project manager as construction
    manager, 135-138

Construction subcontracts, 97-101
Contingency, 78, 272
Contract, administration, 178-183
  overview, 178
  project manager and, 179-181
  thoughts on, 178-179
  typical audit exceptions, 181
Contracting, 83-108, 149-150
  do's and *don't*s of, 100
  engineering services, 101
  general considerations, 84
  overview, 83
  strategy criteria, 87-89
Contracting (sub) construction work,
    97-101
  bid analysis and contract award,
    97- 101
  bidders qualifications, 99-100
  bidding, 100-101
  bid package, 98-99
  overview, 97
Contractor:
  general, 2, 130, 134, 135
  selection, 5
  selection, engineering, procure-
      ment, and construction (EPC)
      contractor, 89-97
    contract award, 96-97
    bidders selection, 93-94
    bid evaluation, 94-97
    bid package, 90-92
    bids, preparation of 94-95
Contracts:
  types of, 84-87
    agency agreement, 84
    competitive, 85
    design/build, 85
    engineering, procurement, and
      construction (EPC), 85
    guaranteed maximum price
      (GMP), 86
    independent contractor agree-
      ment, 84
    lump sum, 85-86

[Contracts]
  negotiated, 85
  reimbursable, 86
  time and material (T&M), 86
  unit price, 86
  typical, 104-108
    agreement, 104
    general terms and conditions,
      105-107
    proposal information, 107
    reimbursable cost schedule, 107-
      108
    scope of work, 104-105
    special terms and conditions,
      107
Control, project (*see* Project control)
Coordination procedure, 53
Cost:
  allocation, 70-77
    buildings, 76
    concrete, 75-76
    dismantling, 77
    electrical, 73-74
    equipment, 72
    field costs, 77
    fire protection, 75
    general, 70-72
    home office, 77
    instrumentation, 73
    insulation, 72
    painting, 76
    piping, 72-73
    site preparation, 74-75
    start up, 77
    structural steel, 76
  control, 143-145
  estimate checking, 79-82
  estimate summary, 67

Design:
  conceptual, 23-25
  detailed, 113-115
  process, 22-35
Design/build contract, 85

Detailed engineering
(*see* Engineering, detailed)
Detailed estimates, 64
Direct cost, 66, 67, 71, 72
Dismantling account, 77
Documentation checklist, 185-187
Durations:
  construction, 12
  project specific, 43
  total project, 18

Electrical account, 73-74
Electric heat tracing, unit prices and
    work hours, 228-229
Engineering, detailed, 6, 109-119
  executed by contractors, 111-116
    basic, 112-113
    coordination and control, 115-
      116
    detailed design, 113-115
  small project execution options,
    116-119
    contracted, 118-119
    in-house, 117-118
Engineering hours, estimating (*see*
    Estimating procedures)
Engineering, procurement and con-
    struction contracts (EPC), 85
  contractor selection, 89-97
Equipment
  account, 72
  estimating procedures, 193-207
    erection, 205-207
    miscellaneous equipment, 202-
      205
    pumps, 199-200
    shell and tube heat exchangers,
      200-201
    vessels, 195-199
Escalation, 78, 272
Estimate:
  adjustments, 77-79
    contingency, 78-79, 272-273
    escalation, 78-272

[Estimate]
    resolution allowance, 77, 271
  anatomy of, 65-69
  breakdown, 69-70
  checking criteria and guidelines,
    79-82
  cost allocation, 70-77
  definitions
    conceptual, 58, 189-190
    definitive, 59, 194
    engineering, 59
    order of magnitude, 58, 189-
      190
    preliminary, 59, 194
Estimating, 5, 55-86
  methods, 59-65
    computerized simulations, 64
    detailed, 64
    factored, 60
    proportioned, 59-60
    semi-detailed, 65, 188-279
    thoughts on, 55-58
    typical factors, 61-63
      fluid plants, 61
      fluid/solids plant, 62
      solids plant, 63
Estimating procedures
  civil work, 200-220
    concrete, 208-214
    miscellaneous civil work, 217-
      220
    structural steel, 214-217
  conceptual estimating, 274-279
  contingency, 272
  electrical, 233-240
  engineering hours, 245-263
  equipment, 193-207
  field costs, 263-266
  instrumentation, 241-245
  insulation, 230-233
  piping, 220-230
  quick, 274-279
    equipment-related costs, 279
    piping, 277-278

Estimating system performance,
    246
Execution plan (*see* Project, execu-
    tion plan/master project
    schedule)
Expediting, 126

Factor, lang, 60
Factored estimates, 60
Factors affecting productivity (*see*
    Labor, productivity)
Factors, typical, 61-63
    all fluid plants, 61
    all solids plant, 63
    fluid/solids plant, 62
Fast track, 20 (*see also* Compressing
    the schedule)
Field:
    costs, 77, 263-271
    indirects, 266-267
    procedures, 136-137
    reports/logs, recommended, 138-
        139
    security system, 137-138
Fire protection:
    account, 75
    estimating, 218
Forecasts, cost and schedule, 170-
    171
Foundations:
    estimating 208-214
    takeoffs guidelines, 209-214

General contractor, 2
    project manager as, 119-121
Guaranteed maximum price contract,
    84

Heat exchangers estimating, 200-
    201
Heat tracing estimating units, 228
    electric, 228
    steam, 228

Independent contractor agreement,
    84
Indirect costs, 66,67, 71, 77
In-house:
    engineering, 117-118
    progress monitoring systems (*see*
        Progress monitoring)
Initial:
    involvement, 8-9
    plan of action, 9-20
        case study, 16-21
        contents, 14-15
        example, 13-14
Inspection, equipment, 127
Instrumentation:
    account, 73
    estimating procedure, 241-245
Insulation:
    account, 76
    estimating procedure, 230-233

Labor:
    cost, 263-265
    productivity, 267-271
    rates, 264
Lang factor, 10, 60
Layout guidelines, plant (*see* Plant
    layout guidelines)
Lump sum contract, 85-86

Major project, 2
Master project schedule (*see*
    Schedule, master project)

Negotiated contract, 85

Options:
    construction, 89, 130-131
    engineering, 88
    small project execution, 116-117
Owner, 1

Painting account, 76
Phase 0/Phase 1, 4, 147

Piping:
  account, 72
  estimating, 220-230
    basic labor units, 228
    carbon steel, 221
    kynar lined, 224
    miscellaneous items
    model, 225
    saran lined, 224
    stainless steel, 304, 316, 222-223
    teflon lined, 224
Planning rules of thumb, 15-16
  construction hours, 15
  engineering hours, 15
  engineering lead time, 16
  equipment count "growth," 15
  peak staff, 16
  process design hours, 15
  total project duration, 15
Plan of action, initial (*see* Initial,
    plan of action)
Plant layout, 151
  guidelines, conceptual, 31-34
    clearances, minimum, 32-34
    maintenance considerations, 32
    safety considerations, 31-32
    typical dimensions, 34
Procedures, field, 136-137
Process design, 4, 22-30
  conceptual, 23
  packages, 23-28
  Phase 1, 26
  Phase 0, 25
  project manager's role in, 28-30
    cost optimization, 28
    Phase 1 review, 29
    Phase 1 specifications, 29-30
Procurement, 6, 122-128
  expediting and inspection criteria,
    125-128
  guideline for purchasing, 123-125
Productivity (*see* Labor, productivity)
Progress monitoring, in-house, 159-
    170

[Progress monitoring, in-house]
  construction, 159-165
  engineering, 165-170
    detailed system, 165-168
    quick system, 168-170
Project:
  control, 7, 141-147
    anatomy of system, 156-158
    construction, during, 154-155
    early stages, in the, 145-150
    engineering office, in the, 150-
      154
    problems, anticipating/ cor-
      recting, 173-176
    project control, during, 154-155
    project manager and the, 143-
      145
    thoughts on, 141-143
  execution plan/master project
    schedule (MPS), 36-54, 148
    case study, 41
    firm, 47
    preliminary, 40-43
    preparation guidelines 39-50
    presentation, 48
  major, 2
  small, 2
Project manager, 1
  as contract administrator, 179-181
  construction manager, 135- 137
  cost control and the, 143-145
  as general contractor, 119-121
  role in Phase 0/Phase 1, 28-30
Proportioned estimate, 59-60
Proposal information, 107
Pumps estimating, 199-200

Quick estimating checks, 274-279
  equipment related, 279
  piping, 277-278
  total installed cost (TIC), 276

Reactors estimating (*see* Vessels
    estimating)

Reimbursable:
  contract, 86
  cost schedule, 107-108
Reports/logs, field, 138-139
Resolution allowance, 77, 271

Scenario, 1
Schedule, 162
  compressing, 50
  control, 145
  master, 5
  master project, 44-47
Scheduling:
  guideline, 42
  influential factors, 38-39
Scope of work, 104-105
Security system, field, 137-138
Semi-detailed estimate, 5, 65, 188-
    279
  procedure, 188-193
Site:
  preparation accounts, 74-75
  selection, 146
  work estimating, 217-218
Small project, 2-4
  execution options, 116-119
Specifications:
  general, 30
  Phase 1, 29
  project, 30
Staffing, peak:
  engineering, 14
  field, 12
Steam tracing, unit prices and work
    hours, 228-229
Structural steel:
  account, 76
  estimating procedure, 214-217
  takeoffs, 216

[Structural steel]
  unit prices/unit hours, 215-216
Tanks estimating (see Vessels esti-
    mating)
Terms and conditions, 105-107
  general, 105-107
  special, 107
Testing, equipment, 128
Time and materials (T&M) contracts,
    86
Tracking curve, 162

Unit price contracts, 86
Unit prices and/or unit hours:
  concrete, 208-209
  electrical work, 239
  equipment erection, 206-207
  instrumentation, 243-244
  insulation, 232-235
  miscellaneous civil work, 217-220
    buildings, 219
    fire protection, 218
    sewers, 218
    site work, 217
  piping, 221-224, 225-226, 229
  structural steel, 215-216
  valves, 231

Value system, 160-162
Valves, unit prices, 231
Venture manager, 1
Vessels estimating, 195-199
  columns, 195-199
  reactors, 195-197
  tanks, 195-197

Work:
  sampling, 176-177
  units, 161